A Global Research and Action Agenda on Cities and Climate Change Science

Cathy Oke · Benjamin Jance IV ·
Gian Carlo Delgado-Ramos ·
Debra C. Roberts · Nicola Tollin ·
Andrew Irvin · Marita Doak
Editors

A Global Research and Action Agenda on Cities and Climate Change Science

The Innovate4Cities Initiative

palgrave
macmillan

Editors
Cathy Oke
Melbourne Centre for Cities, Faculty
Architecture Building and Planning
University of Melbourne
Parkville, VIC, Australia

Gian Carlo Delgado-Ramos
Institute of Geography
National Autonomous University of Mexico
Alcaldía Benito Juárez, Estado de México,
Mexico

Nicola Tollin
Technology and Innovation
University of Southern Denmark
Odense M, Denmark

Marita Doak
Melbourne Centre for Cities
University of Melbourne
Parkville, VIC, Australia

Benjamin Jance IV
Global Covenant of Mayors for Climate
and Energy
Brussels, Belgium

Debra C. Roberts
School of Life Sciences
University of KwaZulu-Natal
Durban, South Africa

Andrew Irvin
Melbourne Centre for Cities
University of Melbourne
Parkville, VIC, Australia

ISBN 978-981-95-6516-0 ISBN 978-981-95-6517-7 (eBook)
https://doi.org/10.1007/978-981-95-6517-7

This Palgrave Macmillan imprint is published by the registered company Springer Nature Singapore Pte Ltd.
The registered company address is: 152 Beach Road, #21-01/04 Gateway East, Singapore 189721, Singapore

If disposing of this product, please recycle the paper.

Foreword

The more than 13,800 cities and local governments in the Global Covenant of Mayors are together making great progress both in cutting emissions and protecting communities from the worsening effects of climate change. Accelerating that progress requires policies that are grounded in science—and strengthened by partnership. That's the focus of the Global Covenant of Mayors' Innovate4Cities initiative, and of the work showcased in this book.

The book illustrates how collaboration between scientists and policy-makers—and across all levels of government—can help us build safer and more sustainable cities while also driving economic growth and expanding opportunity. It explores how Innovate4Cities convenes and advances shared progress, supporting global processes like the IPCC Special Report on Climate Change and Cities (SR Cities) as well as the Coalition for High Ambition Multilevel Partnerships (CHAMP) while empowering local governments with the knowledge and tools needed to take action.

As we enter the next decade of city climate leadership, this volume serves not only as a record of progress, but a blueprint for the future.

Anne Hidalgo
Global Ambassador,
Global Covenant of Mayors for
Climate & Energy (GCoM),
Brussels, Belgium

Andy Deacon
Co-Managing Director,
Global Covenant of Mayors for
Climate & Energy (GCoM),
Brussels, Belgium

Acknowledgments

The editorial team would like to thank all who have been involved in the development of the Innovate4Cities initiative and the Global Research and Action Agenda for Cities and Climate Change Science (GRAA) since the inception of the Conference series at the 2018 Cities and Climate Change Science Conference, held in Edmonton, Canada, and the initiative in 2019. We also thank all colleagues at the Global Covenant of Mayors for Climate and Energy and the Melbourne Centre for Cities at The University of Melbourne, who have assisted the editorial team with reviews, illustrations, figures, and other items for this book.

The editors acknowledge the contributions of all committee members, report writers, and participants in the Cities and Climate Change Science (2018) and Innovate4Cities conferences from 2021 to 2026, including the secretariat and regional offices of: Global Covenant of Mayors for Climate and Energy (GCoM), ICLEI—Local Governments for Sustainability, European Commission—Joint Research Centre (JRC), C40 Cities, United Cities and Local Governments (UCLG), UN-Habitat, European Covenant of Mayors for Climate and Energy, and the many city practitioners, scientists, researchers, and policymakers who have shaped the Global Research and Action Agenda for Cities and Climate Change Science.

Special acknowledgment goes to Nicky Genov for the design of the newest Innovate4Cities graphics throughout the book, to Melanija Tacconi and to copyeditor Laura E. Goodin.

About This Book

A Global Research and Action Agenda on Cities and Climate Change Science: Innovate4Cities explores the dynamic relationships, knowledge gaps, and innovation priorities at the interface of urban practice, transformation, and the science of cities and climate change. Beginning with the landmark 2018 International Conference on Climate Change and Cities in Edmonton and moving through the Innovate4Cities Conferences in 2021 and 2024, this book draws on the uniquely diverse Innovate4Cities global community of policy, research, and practice leaders to trace the development of critical climate change scientific and policy frameworks.

The book presents case studies of city innovation captured in the development and evolution of the Global Research and Action Agenda for Cities and Climate Change Science (GRAA), documenting how this premier conference series has shaped knowledge mobilities, communities of practice, and the global research agenda for the science of cities and climate change.

This Open Access book serves as both a historical record and a forward-looking guide for urban climate action.

Book Structure

The book is organized into three main sections following an introductory chapter:

The introduction establishes cities as key climate actors and frames the multi-dimensional urban characteristics that shape climate responses.

Part 1: *Cities and Climate Change* presents the drivers—people, perspectives, issues—behind the GRAA and associated Conference series.

Part 2: *City Innovation Knowledge Gaps and Action Priorities* examines three structural components for city-level models and data: delivery approaches, pillars of justice and equity, and digitalization.

Part 3: *Systems Approaches to City Research and Innovation* explores interconnected urban systems, including infrastructure, biodiversity, and finance.

The conclusion synthesizes insights and maps a path forward for climate-resilient urban development.

Contents

List of Contributors

Johanna Granados Alcala Global Covenant of Mayors, Brussels, Belgium

Marcus Arcanjo C40 Cities, London, UK

Asad Asadzadeh Institute of Geodesy and Geoinformation, University of Bonn, Bonn, Germany

Zahra Assarkhaniki Faculty of Spatial Sciences, University of Groningen, Groningen, The Netherlands

Juan Pablo Astolfo Global Covenant of Mayors, Brussels, Belgium

Bernhard Barth Global Solutions Division, UN-Habitat, Nairobi, Kenya

Ashali Bhandari Mumbai, India

Eugenie L. Birch University of Pennsylvania, Philadelphia, PA, USA

Hilda Blanco University of Washington, Seattle, WA, USA

Steven Bland World Resources Institute, London, UK

Emily Bolo University of Nairobi, Nairobi, Kenya

Aphrodite Bouikidis Independent (Urban Resilience Specialist), Athens, Greece

Jazmin Burgess London, UK

Judy Bush Faculty of Architecture Building and Planning, University of Melbourne, Melbourne, Australia

Krisztina Campbell London, UK

Agathe Cavicchioli Global Covenant of Mayors for Climate and Energy, Brussels, Belgium

Tanya Chung-Tiam-Fook 7GenCities, Toronto, ON, Canada

Marcus J. Collier Trinity College Dublin, University College Dublin, Dublin, Ireland

David Corbett ClimateView, Stockholm, Sweden

Felix Creutzig Potsdam Institute for Climate Impact Research, Potsdam, Germany

Andy Deacon Global Covenant of Mayors for Climate and Energy, Brussels, Belgium

Gian Carlo Delgado-Ramos Institute of Geography, National Autonomous University of Mexico, Mexico City, Mexico

Marita Doak Melbourne Centre for Cities, Faculty of Architecture Building and Planning, University of Melbourne, Melbourne, Australia

Maria Dombrov Urban Climate Change Research Network, Columbia University Climate School, New York City, NY, USA;
NASA Goddard Institute for Space Studies, New York City, NY, USA

Jayne Engle School of Urban Planning & Max Bell School of Public Policy, McGill University, Montréal, Canada

Niki Frantzeskaki Utrecht University, Utrecht, The Netherlands

Carlos Alberto González-Guzmán MetroCali S.A., Cali, Colombia

Stelios Grafakos Global Green Growth Institute, Budapest, Hungary

Johanna Granados Alcala Global Covenant of Mayors for Climate and Energy, Brussels, Belgium

Julie Greenwalt Cities Alliance, Brussels, Belgium

Kelly Grigsby Municipal Association of Victoria, Melbourne, Australia

Perrine Hamel Asian School of the Environment, Nanyang Technological University, Singapore, Singapore

Angel Hsu University of North Carolina at Chapel Hill, Chapel Hill, NC, USA

Andrew Irvin Melbourne Centre for Cities, Faculty of Architecture Building and Planning, University of Melbourne, Melbourne, Australia

Benjamin Jance IV Global Covenant of Mayors for Climate and Energy, Brussels, Belgium

Asma Jhina Global Covenant of Mayors for Climate and Energy, Brussels, Belgium;
Global Covenant of Mayors, London, UK

Dave Mwita Jomo Ufanisi Research Network, Urban Green Consultants, Nairobi, Kenya

Amy Jones ICLEI— Local Governments for Sustainability, Bonn, Germany;
London School of Economics and Political Science, London, UK

Smruti Jukur Society for the Promotion of Area Resource Centres, Mumbai, India

Susannah Smetana Kagan Urban Climate Change Research Network, Columbia University Climate School, New York City, NY, USA

Aikaterini Karapantsiou Deutsche Gesellschaft für Internationale Zusammenarbeit (GIZ), Bonn, Germany

Bridget Keane Faculty of Architecture Building and Planning, University of Melbourne, Melbourne, Australia

Dave Kendal Future in Nature, Melbourne, Australia

Samia Khan Global Covenant of Mayors for Climate and Energy (GCoM), Brussels, Belgium

Anna Koch Global Covenant of Mayors for Climate and Energy, Brussels, Belgium

Theo Kötter University of Bonn, Bonn, Germany

Benjamin Leffel University of Nevada Las Vegas, Las Vegas, NV, USA

Mattia Federico Leone Department of Architecture, University Naples Federico II, Naples, Italy

Florence Lozet Cities Alliance, Brussels, Belgium

Pooja Mahapatra Fugro, The Randstad, The Netherlands

Clara Marais ICLEI Africa, Cape Town, South Africa

Pablo Mariani United Cities and Local Governments, Brussels, Belgium

Joanna Masic World Bank, Washington, DC, USA

Timon McPhearson Urban Systems Lab, New York University, New York, NY, USA;
Cary Institute of Ecosystem Studies, Milllbrook, NY, USA;
Beijer Institute of Ecological Economics at the Swedish Royal Academy of Sciences, Stockholm, Sweden

Marnie McGregor Global Covenant of Mayors for Climate and Energy (GCoM), Brussels, Belgium

Giulia Melica European Commission, Joint Research Centre (JRC), Ispra, Italy

David Miller C40 Centre for City Climate Policy and Economy, Toronto, ON, Canada

Sareh Moosavi Faculty of Architecture Building and Planning, University of Melbourne, Melbourne, Australia

Ana Rosa Moreno Sánchez Faculty of Medicine, National Autonomous University of Mexico (Retired), Mexico City, Mexico

Carlos Moreno IAE Paris Sorbonne, Université Paris 1 Panthéon-Sorbonne, Paris, France

Priscilla Negreiros Cities Climate Finance Leadership Alliance, London, UK

Ben Odongo ICLEI Africa, Cape Town, South Africa

Cathy Oke Melbourne Centre for Cities, Faculty of Architecture Building and Planning, University of Melbourne, Melbourne, Australia

Juliet Oluoch Ufanisi Research Network, Nairobi, Kenya

Juliet Aluoch Oluoch Ufanisi Research Network, Nairobi, Kenya

Valentina Palermo European Commission, Joint Research Centre (JRC), Ispra, Italy

Dennis Pamlin RISE Research Institute of Sweden, Gothenburg, Sweden

Sheela Patel Society for the Promotion of Area Resource Centres, Mumbai, India

Daniel Pejic Melbourne Centre for Cities, University of Melbourne, Melbourne, Australia

Melissa Pineda-Pinto Faculty of Architecture Building and Planning, University of Melbourne, Melbourne, Australia

Giorgia Rambelli Global Covenant of Mayors for Climate and Energy (GCoM), Brussels, Belgium;
Urban Transitions Mission (UTM), Global Covenant of Mayors for Climate and Energy (GCoM), Brussels, Belgium

Lea Ranalder UN-Habitat, Nairobi, Kenya

Mauricio Rodas University of Pennsylvania, Philadelphia, PA, USA

Debra C. Roberts School of Life Sciences, University of KwaZulu-Natal, Durban, South Africa;
Faculty of Geo-Information Science and Earth Observation, University of Twente, Enschede, The Netherlands

Paty Romero-Lankao Department of Sociology, University of Toronto Scarborough, Scarborough, Canada

Cynthia Rosenzweig Urban Climate Change Research Network, Columbia University Climate School, New York City, NY, USA;
NASA Goddard Institute for Space Studies, New York City, NY, USA

Natalia Salazar ICLEI—Local Governments for Sustainability, Bonn, Germany

Pourya Salehi ICLEI—Local Governments for Sustainability, Bonn, Germany

Simone Sandholz Institute for Environment and Human Security, United Nations University, Cologne, Germany

Caterina Sarfatti Milan, Italy

Asif Nawaz Shah Global Covenant of Mayors for Climate and Energy, Brussels, Belgium

William Solecki Urban Climate Change Research Network, Columbia University Climate School, New York City, NY, USA;
Department of Geography and Environmental Science, Urban Climate Change Research Network, Hunter College—CUNY, New York City, NY, USA

Meggan Spires ICLEI Africa, Cape Town, South Africa

Nicola Tollin UNESCO Chair on Urban Resilience, University of Southern Denmark, Odense, Denmark

Ika Trijsburg Crawford School of Public Policy, Australian National University, Canberra, Australia

Maryke van Staden Carbonn Climate Center, ICLEI Local Governments for Sustainability, Bonn, Germany

Jorn Verbeeck Dark Matter Labs, Amsterdam, The Netherlands;
Dark Matter Labs, Antwerp, Belgium

Martin Wainsten OpenEarth Foundation, San Diego, CA, USA

Brenna Walsh Vancouver, Canada

List of Figures

List of Tables

CHAPTER 1

Introduction: Cities at the Forefront of Climate Change

Benjamin Jance IV, Cathy Oke, Amy Jones, Bernhard Barth, Kelly Grigsby, Andrew Irvin, Nicola Tollin, and Marita Doak

1.1 Introduction

The emergence, growth, and evolution of cities—from the smallest to the largest—have been driven by the needs of the people who inhabit them. More than 6000 years ago, the world's first "cities"—agglomerations of people and goods within a concentrated space—began to emerge in ancient Mesopotamia, buoyed by the advent of farming, the surplus production of food, and early prospects for trade (Yoffee &

B. Jance IV (✉)
Global Covenant of Mayors for Climate and Energy, Brussels, Belgium
e-mail: bjance@globalcovenantofmayors.org

C. Oke · A. Irvin · M. Doak
Melbourne Centre for Cities, Faculty of Architecture Building and Planning, University of Melbourne, Melbourne, Australia
e-mail: cathy.oke@unimelb.edu.au

C. Oke et al. (eds.), *A Global Research and Action Agenda on Cities and Climate Change Science*,
https://doi.org/10.1007/978-981-95-6517-7_1

Terrenato, 2015). More than 250 years ago, cities again transformed into engine rooms of industry, production, and rapid modernization through the Industrial Revolution, sparking economic growth, geopolitical shifts, and changes to the relationship between humans and their surrounding environments (Brunt et al., 2021). Now housing over half the world's population, urban areas generate three-quarters of global economic output and account for 67–72% of global greenhouse gas emissions—a share that continues to rise as urban populations expand by over 400 million inhabitants every five years (Lwasa et al., 2022; Vogel et al., 2024). While cities today face the myriad interconnected needs of billions of people, two particular needs stand out as both necessities and opportunities that could define the next era of how urban areas look, function, and live: the need to rapidly reduce greenhouse gas (GHG) emissions that contribute to climate change, and the need to adapt to the worsening risks and hazards that accompany it. Together, these needs comprise the underlying *raison d'etre* of this book, which offers challenges, solutions,

A. Irvin
e-mail: irvin.a@unimelb.edu.au

M. Doak
e-mail: marita.doak@unimelb.edu.au

A. Jones
London School of Economics and Political Science, London, UK
e-mail: a.l.jones2@lse.ac.uk

B. Barth
Global Solutions Division, UN-Habitat, Nairobi, Kenya
e-mail: bernhard.barth@un.org

K. Grigsby
Municipal Association of Victoria, Melbourne, Australia
e-mail: kgrigsby@mav.asn.au

N. Tollin
UNESCO Chair on Urban Resilience, University of Southern Denmark, Odense, Denmark
e-mail: nto@iti.sdu.dk

and opportunities to once again take advantage of the ingenuity, collaboration, and scale of cities to resoundingly respond to the threats of climate change.

With 2.5 billion people projected to join urban populations by 2050, concentrated predominantly in Asia and Africa, cities represent an unprecedented opportunity for high-impact climate and energy interventions (UN DESA, 2018; UN-Habitat, 2024; UNESCAP, 2020). These actions—as further chapters of this book will illustrate, vary widely in scope, scale, and effectiveness—have arisen out of an inherent need to respond in the face of danger. Cities confront increasingly intense and frequent climate risks and hazards, the impacts of which endanger the lives of millions (GCoM, 2024; Goldhill & Fitzgibbon, 2021). Flooding and extreme precipitation, extreme heat, wildfires, and water scarcity—the negative impacts of which are amplified by urban heat islands and impervious surfaces—represent more than 60% of the climate hazards that GCoM signatories reported in 2024 (Cheval et al., 2024; Dharmarathne et al., 2024; GCoM, 2024). Vulnerable populations, including women, children, and those living in informal settlements, experience disproportionate risks, while critical infrastructure faces unprecedented threats from hydrological extremes and temperature increases (GCoM, 2025; IPCC, 2022; Le et al., 2021). Evidence increasingly suggests that the risks and impacts wrought by these climate hazards are set to escalate beyond those described in a 1.5 °C scenario (Kapucu et al., 2024). This is compounded by the emerging projections that humanity is tracking to "overshoot" the Paris Agreement's commitment to pursue keeping global temperature rise to 1.5 °C, carrying a high probability of deleterious effects for urban environments, and particularly for the most vulnerable people who inhabit them from both the exceedance and return to that temperature limit (Bauer et al., 2023; Gasser et al., 2015; Peters, 2024; Reisinger et al., 2025).

1.1.1 Cities Are Responding With Action

Faced with increasing demographic shifts and intensifying effects of climate change, cities have already begun taking up the mantle of "first responder" and offering a counter-offensive to its associated risks. Cities worldwide have signaled the urgency of action through formal declarations and commitments (Bush & Doyon, 2025; Too et al., 2024). Cities and urban areas function, as they historically have, as laboratories for innovation, offering unparalleled opportunities for integrated adaptation and mitigation strategies that maximize co-benefits across multiple sustainability challenges (Bulkeley & Newell, 2023; Seto et al., 2014). The spatial concentration of at-risk people, infrastructure, and economic activities in urban areas means that while the value of goods and services increases in the face of climate change, economies of scale also enable more-systemic approaches to climate-relevant solutions, supported by urban governance structures that can often rapidly implement transformative policies across all city sizes and development levels (Moosavi et al., 2023; O'Neill et al., 2023). The Paris Agreement—the mutually binding document produced at the 21st UN Climate Change Conference (COP21) that commits parties to the United Nations Framework Convention on Climate Change (UNFCCC) to keeping global temperature increase to "well below 2 °C" and "pursuing efforts to limit it to 1.5 °C"—highlights the role of subnational and local levels as critical opportunities for implementation (UNFCCC, 2015).

Recent analysis also reveals that 78% of the 194 Nationally Determined Contributions submitted as of 2025 (up from 66% in 2023) contain moderate or strong urban content, signaling that national governments are beginning to formally view cities and local governments as partners in climate-action implementation—although the extent to which these city contributions are formally integrated into national NDCs varies across countries (Ranalder et al., 2025). This increasing recognition of cities within national climate frameworks is complemented by bottom-up urban initiatives that demonstrate the power of collective city-level climate action. Over 13,700 cities, representing a combined total of over 1 billion inhabitants, have joined the Global Covenant of Mayors for Climate and Energy (GCoM), the world's largest alliance of cities and local governments committed to climate action (GCoM, 2024).

Although further structured efforts are required at the national level to recognize and strengthen the inclusion and prioritization of cities within national climate policies—including NDCs, NAPs, and LT-LEDSs—a stronger and broader urban focus can enhance the ambition of national climate policies and support the achievement of national climate goals (UN-Habitat et al., 2024). Such an approach can also facilitate the provision of financial, technological, and institutional support to subnational governments. To include cities in NDCs and other climate policies, it is essential to establish clear baselines for GHG emissions and climate risks, including vulnerabilities and exposures; identify challenges and corresponding responses for both mitigation and adaptation; and assess urban needs for finance, technology, and capacity-building (Ranalder et al., 2025).

Analysis of data from the over 4500 GCoM signatory cities that have submitted greenhouse gas inventories reveals that these cities account for approximately 10 gigatons (Gt) of GHG emissions. To contextualize this impact: achieving Paris-aligned emissions reductions across this subset of GCoM cities alone would reduce global emissions by approximately 1.8 Gt CO_2e—equivalent to removing all of Japan's emissions (C40 & GCoM, 2025). Cities have set short-term targets that, if achieved, would close 37% of the gap between current NDCs and Paris-aligned emissions-reduction trajectories (C40 & GCoM, 2025). The innovative approaches proliferating across cities range from nature-based solutions to technological innovations, governance reforms, and finance mechanisms, as evidenced by initiatives like Copenhagen's integrated district heating system (City of Copenhagen, 2021), Freetown's #FreetownTheTreeTown urban-forestry campaign (City of Freetown, 2023), and Barcelona's energy-poverty strategy (City of Barcelona, 2023). These examples demonstrate that cities are not merely responding to climate change but actively shaping the global climate agenda through concrete actions that combine mitigation ambition with adaptation necessity.

1.1.2 Climate Action in Cities Requires Comprehensive Support

Despite this groundswell of action and support, critical gaps remain if cities are to progress in their "journeys" toward low-emission, climate-resilient places and spaces. Carbon Disclosure Project (CDP)'s 2025 study of 611 cities across 75 countries highlights that they require US$86 billion in investment to achieve their stated adaptation and resilience goals (CDP, 2025). This investment need reflects a broader infrastructure-financing challenge. Enabling climate action in cities demands transformative shifts across multiple domains: financial systems, governance structures, capacity-building, technology deployment, and robust data collection and monitoring infrastructures (Acuto & Leffel, 2020; Bulkeley, 2009; Bulkeley & Betsill, 2013; Bulkeley & Castán Broto, 2013).

Globally, the modeled investment requirements for 2020 to 2030 in scenarios that limit warming to 2 °C or 1.5 °C are a factor of three to six greater than current levels, with the largest gaps in developing countries where rapid urbanization occurs, creating challenges for cities in these regions (IPCC, 2022, 2023). The magnitude of this gap underscores the urgent need for scaled-up climate-finance mechanisms that can channel resources on an urban scale. The resources for adaptation present an even starker picture. The adaptation-financing gap—estimated at US$310–366 billion annually—noted at 10 to 18 times larger than current international public adaptation-finance flows, and is not reducing (UNEP, 2023, 2025). This disparity is especially acute in urban contexts where adaptation investments must address multiple, intersecting vulnerabilities.

In addition to the challenges associated with finance, comprehensive adaptation strategies still prove insufficient. Current and planned adaptation measures cannot resolve most urban climate risks, creating persistent adaptation gaps across all regions and hazard types. Urban dwellers face immediate climate impacts that are already locked in due to historical emissions, as well as escalating future risks that will require unprecedented adaptation responses. Even with ambitious mitigation efforts, cities must prepare for unavoidable changes including sea-level rise, more-frequent extreme heat events, and intensified precipitation patterns that will reshape urban life and require transformative adaptation. Technology transfer and capacity-building in developing countries' cities lag where there are critical needs, particularly in accessing cutting-edge climate technologies and adaptation knowledge (IPCC, 2022). Developing countries' cities face the dual challenge of addressing historical infrastructure deficits while simultaneously building climate resilience into rapidly expanding urban footprints, all within constrained fiscal environments and often without adequate institutional frameworks or technical expertise.

Beyond financing and infrastructure, successful urban climate action depends on integrated governance frameworks that embed climate considerations across all municipal operations from transportation and energy systems to housing, water management, and public health (Bulkeley & Betsill, 2013; Medeiros & van der Zwet, 2020). This integration requires sophisticated data systems, robust monitoring and evaluation frameworks, and sustained political commitment that often extends beyond electoral cycles. Cities also need mechanisms for meaningful community engagement that ensure climate actions address the needs of the most vulnerable populations and contribute to just transitions that do not exacerbate existing inequalities (Anguelovski et al., 2016).

1.2 The Origins, Evolution, and Impact of the Innovate4Cities Initiative and the Global Research and Action Agenda

The 2018 Cities and Climate Change Science Conference, held in Edmonton, Canada, marked a significant shift in integrating urban perspectives into global climate science. This landmark conference, attended by over 700 participants from diverse urban stakeholders including policymakers, academics, innovators, and NGOs, fostered a transdisciplinary research approach to collaboratively establish research priorities aimed at enhancing urban climate resilience. The conference, which was co-sponsored by the IPCC, led directly to the creation of the first Global Research and Action Agenda on Cities and Climate Change Science (GRAA 2018) (Prieur-Richard et al., 2018). Introduced at the IPCC's 48th Plenary in Incheon, Republic of Korea, in 2018, the GRAA serves as a comprehensive guide for future research and practical actions on urban climate challenges, with particular emphasis on urban informality, a key novel contribution identified during the Edmonton event (Prieur-Richard et al., 2019).

Building directly upon the 2018 conference outcomes, the Innovate4Cities initiative was established to further expand these collaborative efforts, underscoring the integration of scientific knowledge, policymaking, and innovative practices to accelerate urban climate action. Innovate4Cities explicitly highlights multi-level governance, strategic partnerships, digitalization, and finance as critical themes necessary for advancing urban resilience and sustainability (Hunter et al., 2023). The initiative has continued to convene stakeholders through subsequent conferences, with each iteration updating and refining the GRAA to reflect evolving urban climate challenges and opportunities. Most recently, at the 2024 Innovate4Cities Conference in Montréal, Canada, the GRAA was updated to incorporate new research priorities and action pathways that reflected six years of implementation experience, emerging climate science, and lessons learned from cities worldwide (Irvin et al., 2024). This latest edition, GRAA 2024, maintains the systems approach to research and innovation in urban planning and design, using city-level models to deliver just and equitable action in climate-resilient development. The 2024 GRAA highlights that achieving transformative urban climate action requires not only technical innovations but also fundamental shifts in governance structures, financial mechanisms, and knowledge-exchange practices.

The implementation phase following the 2018 conference has underscored the pivotal role played by individual knowledge brokers and informal networks in mobilizing urban climate knowledge and policy across various scales. This mobilization has significantly enhanced the capacity of cities to integrate scientific research into urban planning and governance frameworks, facilitating more robust local responses to climate challenges and effectively translating global climate science into local urban solutions (Hunter et al., 2023). The Innovate4Cities initiative is actively supporting authors in the ongoing development of the IPCC Special Report on Climate Change and Cities (SRCCC), scheduled for release in 2027 as part of the IPCC's Seventh Assessment Report (AR7) cycle, demonstrating the direct influence of this transdisciplinary approach on global climate assessment processes.

The GRAA and Innovate4Cities initiative have identified several critical enabling conditions that cities require to translate research and commitments into effective climate action. These enabling conditions—spanning governance structures, data systems, infrastructure planning, and community engagement—represent the foundational elements necessary for cities to implement the ambitious climate strategies outlined in their plans. Understanding these requirements is essential to supporting cities in their climate journeys.

First among these enabling conditions is the embedding of climate considerations across all municipal departments. Rather than treating climate action as isolated initiatives confined to environmental or sustainability offices, successful cities integrate climate objectives into transportation planning, energy systems, housing development, water management, public health programs, and economic development strategies (Bulkeley & Betsill, 2013). This horizontal integration ensures that climate considerations inform decision-making at every level and prevent disconnected approaches that can undermine overall climate goals. Such integration requires strong political leadership, clear institutional mandates, and mechanisms for cross-departmental coordination that overcome traditional bureaucratic barriers.

Second, robust data collection and monitoring systems form the backbone of evidence-based urban climate action. Cities need comprehensive greenhouse gas inventories, climate risk assessments, and vulnerability mapping to establish baselines and identify priorities. Advanced data systems—including smart city sensors, satellite monitoring, and digital twins—can provide real-time information that enables adaptive management and early-warning systems (Kitchin et al., 2015; Ricciardi & Callegari, 2022). However, data collection alone has proven insufficient without analytical capacity to transform raw data into actionable insights, and transparent reporting mechanisms that track progress against targets and enable accountability (Rosenzweig et al., 2018). Cities such as Lima, Peru have demonstrated this approach through comprehensive air-quality monitoring networks that inform targeted interventions (Municipality of Lima, 2024).

Third, the physical infrastructure that shapes urban form and function must be reimagined for climate resilience. This encompasses both "gray" infrastructure—roads, buildings, and energy and water systems—and nature-based solutions and ecosystem-based adaptation, such as urban forests, wetlands, and green corridors, that provide ecosystem services while enhancing resilience (Seto et al., 2014). Infrastructure planning must account for climate projections throughout asset lifecycles, incorporate redundancy and flexibility to withstand shocks, and prioritize solutions that deliver multiple co-benefits across mitigation, adaptation, and development goals (Davoudi, 2016). Quezon City's GrowQC urban-agriculture initiative exemplifies how infrastructure interventions can simultaneously address food security, green-space provision, and climate adaptation (Quezon City Government, 2022).

Fourth, meaningful community engagement and participatory governance structures ensure that climate actions reflect local needs and priorities while building social capital and collective efficacy (Cloutier et al., 2014; Fougère & Meriläinen, 2019). Effective participation extends beyond consultation to genuine co-production of knowledge and solutions, which is particularly important in informal settlements and marginalized communities, the residents of which often possess invaluable adaptation knowledge but lack formal platforms for input (Shand, 2017). Co-production approaches can strengthen both the technical quality and social legitimacy of climate interventions while building community ownership that enhances long-term sustainability (Almansi et al., 2011; Joshi & Moore, 2004).

Fifth, vertical integration between local, regional, national, and international governance scales creates enabling policy environments for urban climate action. Cities operate within multi-scalar governance systems where higher-level policies can either facilitate or constrain local initiatives. The CHAMP (Coalition for High Ambition Multilevel Partnership) initiative exemplifies efforts to strengthen these vertical linkages by connecting cities with national governments and international institutions to align climate strategies across scales (ICLEI, 2024). As countries prepare their next generation of Nationally Determined Contributions (NDCs 3.0), opportunities exist to formally integrate city contributions and establish mechanisms for ongoing coordination (UNFCCC, 2025).

1.3 Structure and Scope of the Book

This book is structured to guide readers through the evolution, implementation, and future directions of Innovate4Cities and its knowledge-action interface—the GRAA—which connects research with practice. The chapters are organized into three interconnected parts that build progressively from the GRAA's foundational concepts through its enabling mechanisms to integrated systems approaches.

Section I: Cities and Climate Change establishes the conceptual and institutional frameworks that underpin contemporary urban climate action. Following this introductory chapter, subsequent chapters trace the development of transdisciplinary approaches to urban climate research, examine the evolution of city networks and governance arrangements, and analyze how global climate science has been progressively localized to inform context-specific urban interventions. These foundational chapters demonstrate how cities have emerged as critical actors in climate governance while highlighting persistent challenges in translating global frameworks to diverse local contexts.

Section II: City Innovation Knowledge Gaps and Action Priorities explores the critical systems, capacities, and conditions required for cities to implement effective climate action. Chapters in this section examine finance and investment mechanisms, data infrastructure and monitoring systems, institutional capacity-building, and the role of innovation ecosystems in accelerating climate solutions. Particular attention is devoted to addressing the specific challenges faced by cities in developing countries, where resource constraints, institutional limitations, and rapid urbanization create unique implementation barriers. These chapters provide practical guidance on overcoming common obstacles while showcasing successful approaches from diverse urban contexts.

Section II: Systems Approaches to City Research and Innovation: synthesizes insights from earlier sections to present integrated frameworks for urban climate action that address the interconnected nature of mitigation, adaptation, and development. Chapters examine sector-specific strategies across energy, transportation, buildings, water, and nature-based solutions while emphasizing cross-sectoral coordination and systems thinking.

The concluding chapter looks ahead to emerging challenges and opportunities, including the IPCC Special Report on Climate Change and Cities, pathways for scaling successful innovations, and recommendations for ensuring just and equitable transitions that address historical inequities while building climate resilience.

Throughout the book, case studies from cities across global development contexts illustrate key concepts and demonstrate the diversity of approaches being implemented. These real-world examples highlight both successes and failures, providing valuable lessons for practitioners, policymakers, researchers, and students engaged in urban climate action. The book's emphasis on transdisciplinary collaboration and knowledge co-production reflects the core principles of the GRAA and Innovate4Cities initiative, recognizing that addressing the urban climate challenge requires sustained partnerships that bridge divides between research and practice, Global North and Global South, formal and informal, and technical and social dimensions of transformation.

By documenting the evolution of urban climate change science and action from Edmonton 2018 through Montréal 2024 to Nairobi 2026 and beyond, this book aims to provide both a retrospective analysis of progress achieved and a prospective roadmap for accelerating the transformation required to align urban development with the Paris Agreement's 1.5 °C temperature limit. As cities continue to grow in population, economic importance, and greenhouse gas emissions, the urgency of this transformation intensifies. This book demonstrates that while significant challenges remain, cities possess the creativity, resources, and determination needed to lead the global climate response—provided they receive the enabling support, strategic partnerships, and sustained commitment required to turn ambition into reality.

Acknowledgements The authors would like to thank Veronica Castiglione for further contributions to ensuring the cohesiveness of this chapter.

Competing Interests

The authors have no conflicts of interest to declare that are relevant to the content of this chapter.

References

Acuto, M., & Leffel, B. (2020). Understanding the global ecosystem of city networks. *Urban Studies, 58*(9), 1758–1774.

Almansi, F., Hardoy, J., & Pelling, M. (2011). Urban adaptation to climate change: New challenges for governance. *IIED Human Settlements Working Paper Series, 3*, 1–32.

Anguelovski, I., Shi, L., Chu, E., Gallagher, D., Goh, K., Lamb, Z., Reeve, K., & Teicher, H. (2016). Equity impacts of urban land use planning for climate adaptation: Critical perspectives from the global north and south. *Journal of Planning Education and Research, 36*(3), 333–348.

Bauer, N., Kalkuhl, M., Fesenfeld, L., & Popp, A. (2023). Ratcheting up climate ambition: Challenges and opportunities for implementing the Paris agreement. *Wires Climate Change, 14*(5), Article e829.

Brunt, L., Cannon, E., & Schulze, M. S., et al. (2021). The economic history of Britain's industrial revolution. In R. Floud (Ed.), *The Cambridge economic history of modern Britain* (pp. 1–44). Cambridge University Press.

Bulkeley, H. (2009). *Cities and climate change*. Routledge.

Bulkeley, H., & Betsill, M. (2013). Revisiting the urban politics of climate change. *Environmental Politics, 22*(1), 136–154.

Bulkeley, H., & Castán Broto, V. (2013). Government by experiment? Global cities and the governing of climate change. *Transactions of the Institute of British Geographers, 38*(3), 361–375.

Bulkeley, H., & Newell, P. (2023). *Governing climate change* (3rd ed.). Routledge.

Bush, E., & Doyon, B. (2025). Climate emergency declarations: Urban climate governance and the politics of urgency. *Climate Policy, 25*(2), 145–162.

C40 and GCoM. (2025). *Cities leading climate action: 2025 global analysis*. Global Covenant of Mayors for Climate and Energy.

CDP. (2025). *Cities at risk: Global climate adaptation investment needs*. CDP Worldwide.

Cheval, S., Pîrvulescu, C., & Akbari, H. (2024). Urban heat island effects and mitigation strategies: A global perspective. *Climate, 12*(4), 48.

City of Barcelona. (2023). *Energy poverty strategy 2030*. Retrieved November 4, 2025 from https://www.barcelona.cat/

City of Copenhagen. (2021). *Copenhagen district heating system report*. Retrieved November 4, 2025 from https://www.kk.dk/

City of Freetown. (2023). *#FreetownTheTreeTown campaign report*. Retrieved November 4, 2025 from https://www.freetown.gov.sl/

Cloutier, G., Joerin, F., Dubois, C., Labarthe, M., Legay, C., & Viens, D. (2014). Planning adaptation based on local actors' knowledge and participation: A climate governance experiment. *Planning Practice & Research, 29*(4), 458–474. https://doi.org/10.1080/14693062.2014.937388

Davoudi, S. (2016). The value of planning and the values in planning. *Town Planning Review, 87*(6), 615–618. https://doi.org/10.3828/tpr.2016.40

Dharmarathne, T. S., Weerasinghe, W. M., & Perera, P. (2024). Urban flooding and climate hazards in tropical cities. *Natural Hazards Review, 25*(1), 04023042.

Fougère, M., & Meriläinen, E. (2019). Exposing three dark sides of social innovation through critical perspectives on resilience. *Industry and Innovation, 26*(8), 990–1007. https://doi.org/10.1080/13662716.2019.1709420

Gasser, T., Guivarch, C., Tachiiri, K., Jones, C. D., & Ciais, P. (2015). Negative emissions physically needed to keep global warming below 2 °C. *Nature Communications, 6*, 7958.

GCoM. (2024). *Global state of cities climate action 2024*. Global Covenant of Mayors for Climate and Energy.

GCoM. (2025). *Urban climate risk assessment: 2025 update*. Global Covenant of Mayors for Climate and Energy.

Goldhill, O., & Fitzgibbon, W. (2021). Climate change and urban vulnerability: A global assessment. *Environmental Research Letters, 16*(9), Article 094056.

Hunter, N., Walsh, B., Ramos, G. C. D., Greenwalt, J., Roberts, D., Oke, C., Nielsen, D., & Slotow, R. (2023). Cities and climate change science after Edmonton: Knowledge, people, and organisations' influence and outreach. *Current Research in Environmental Sustainability, 6*, Article 100231.

ICLEI. (2024). *CHAMP: Coalition for high ambition multilevel partnership for climate action*. Retrieved November 4, 2025 from https://iclei.org/champ/

IPCC. (2022). Climate change 2022: Impacts, adaptation and vulnerability. Contribution of Working Group II to the Sixth Assessment Report of the Intergovernmental Panel on Climate Change [H.-O. Pörtner, D. C. Roberts, M. Tignor, E. S. Poloczanska, K. Mintenbeck, A. Alegría, M. Craig, S. Langsdorf, S. Löschke, V. Möller, A. Okem, & B. Rama (Eds.)]. Cambridge University Press. https://doi.org/10.1017/9781009325844

IPCC. (2023). Summary for policymakers. In Core Writing Team, H. Lee, & J. Romero (Eds.), *Climate change 2023: Synthesis report. Contribution of working groups I, II and III to the sixth assessment report of the intergovernmental panel on climate change* (pp. 1–34). IPCC. https://doi.org/10.59327/IPCC/AR6-9789291691647.001

Irvin, A., Oke, C., Khan, S., Jance IV, B., Ranalder, L., Barth, B., Deacon, A., Abusamra, R., Addison, A., Adebisi, J., Adsett, C., Ailio, E., Al Jadaa, D., Aslan, S., Assalini, S., Assarkhaniki, Z., Astolfo, J. P., Badino, M., Bai, X., …, & Williams, A. (2024). *Next-generation city climate action: Findings from the 2024 Innovate4Cities conference and update to the global research and action agenda on cities and climate change science (GRAA)*. Global Covenant of Mayors for Climate and Energy, UN-Habitat & University of Melbourne. https://doi.org/10.26188/27147612

Joshi, A., & Moore, M. (2004). Institutionalised co-production: Unorthodox public service delivery in challenging environments. *Journal of Development Studies, 40*(4), 31–49. https://doi.org/10.1080/00220380410001673184

Kapucu, N., Hawkins, C. V., & Rivera, F. I. (Eds.). (2024). *Disaster resilience and sustainability: Japan's urban development and social capital.* Springer.

Kitchin, R., Lauriault, T. P., & McArdle, G., et al. (2015). Smart cities and the politics of urban data. In S. Marvin (Ed.), *Smart urbanism* (pp. 16–33). Routledge.

Le, T. D. N., Huynh, V. T., & Nguyen, H. Q. (2021). Climate vulnerability of informal settlements: A systematic review. *Urban Climate, 37*, Article 100851.

Lwasa, S., Seto, K. C., Bai, X., Blanco, H., Gurney, K. R., Kılkış, Ş., Lucon, O., Murakami, J., Pan, J., Sharifi, A., & Yamagata, Y. (2022). Urban systems and other settlements. In *IPCC, 2022: Climate change 2022: Mitigation of climate change. Contribution of working group III to the sixth assessment report of the intergovernmental panel on climate change.* Cambridge University Press. https://doi.org/10.1017/9781009157926.010

Medeiros, E., & van der Zwet, A. (2020). Sustainable and integrated urban planning and governance in metropolitan and medium-sized cities. *Sustainability, 12*(15), 5976.

Moosavi, S., Browne, G. R., & Bush, J. (2023). Dynamics of urban climate governance: Examining the role of non-state and subnational actors. *Climate Policy, 23*(3), 356–371.

Municipality of Lima. (2024). *Air quality monitoring network expansion report.* Retrieved November 4, 2025 from https://www.munlima.gob.pe/

O'Neill, B. C., van Aalst, M., Ibrahim, Z. Z., et al. (2023). Key risks across sectors and regions. In H.-O. Pörtner, et al. (Eds.), *Climate change 2023: Impacts, adaptation and vulnerability* (pp. 2411–2538). Cambridge University Press.

Peters, G. P. (2024). Carbon budget implications of temperature overshoot. *Environmental Research Letters, 19*(3), Article 034015.

Prieur-Richard, A. H., Walsh, B., Craig, M., Melamed, M. L., Colbert, M. L., Pathak, M., Connors, S., Bai, X., Barau, A. S., Bulkeley, H., & Cleugh, H. (2019). Global research and action agenda on cities and climate change science. In *Cities & climate change science conference.*

Prieur-Richard, A.-H., Walsh, B., Craig, M., Melamed, M. L., Colbert, M., Pathak, M., Connolly, D., Leonardsen, D., Wright, G., Oke, C., Parnell, J., Okem, A., Hayward, B., Rosenzweig, C., Grimm, N. B., Moss, R., Seto, K. C., & Solecki, W. (2018). *Global research and action agenda on cities and climate change science – Full version.* Retrieved November 4, 2025 from https://www.ipcc.ch/site/assets/uploads/2018/12/GRAA-Cities-and-Climate-Change-Science-Full-Report.pdf

Quezon City Government. (2022). *GrowQC urban agriculture initiative report*. Retrieved November 4, 2025 from https://quezoncity.gov.ph/

Ranalder, L., Tollin, N., Gragnani, P., & Geoghegan, P. (2025). *Seizing the urban opportunity: A 3-step guide for strengthening Nationally Determined Contributions (NDC 3.0)*. United Nations Human Settlements Programme (UN-Habitat). Retrieved November 4, 2025 from https://unhabitat.org/seizing-the-urban-opportunity-a-3-step-guide-for-strengthening-nationally-determined-contributions

Reisinger, A., Fuglestvedt, J. S., Pirani, A., Geden, O., Jones, C. D., Maharaj, S., Poloczanska, E. S., Morelli, A., Johansen, T. G., Adler, C., Betts, R. A., & Seneviratne, S. I. (2025). Overshoot: A conceptual review of exceeding and returning to global warming of 1.5°C. *Annual Review of Environment and Resources, 50*, 185–217. https://doi.org/10.1146/annurev-environ-111523-102029

Ricciardi, G., & Callegari, G. (2022). Digital twins for climate-neutral and resilient cities: State of the art and future development as tools to support urban decision-making. In *International conference on technological imagination in the green and digital transition* (pp. 617–626). Springer.

Rosenzweig, C., Solecki, W. D., Romero-Lankao, P., Mehrotra, S., Dhakal, S., & Ibrahim, S. A. (Eds.). (2018). *Climate change and cities: Second assessment report of the Urban Climate Change Research Network*. Cambridge University Press.

Seto, K. C., Dhakal, S., Bigio, A., Blanco, H., Delgado, G. C., Dewar, D., Huang, L., Inaba, A., Kansal, A., Lwasa, S., & McMahon, J., et al. (2014). Human settlements, infrastructure, and spatial planning. In O. Edenhofer (Ed.), *Climate change 2014: Mitigation of climate change* (pp. 923–1000). Cambridge University Press.

Shand, W. (2017). Efficacy in action: Mobilising community participation for inclusive urban development. *Development in Practice, 27*(8), 1146–1158. https://doi.org/10.1080/09614524.2017.1367099

Too, L., Bajracharya, B., & Hastings, J. (2024). Urban climate emergency declarations: Symbolic politics or substantive action? *Environmental Planning and Management, 67*(3), 542–561.

UN DESA. (2018). *World urbanization prospects: The 2018 revision*. United Nations Department of Economic and Social Affairs.

UNEP. (2023). *Adaptation gap report 2023: Underfinanced, underprepared*. United Nations Environment Programme.

UNEP. (2025). *Adaptation gap report 2025: Running on empty. The world is gearing up for climate resilience—Without the money to get there*. United Nations Environment Programme. https://doi.org/10.59117/20.500.11822/48798

UNESCAP. (2020). *The future of Asian & Pacific cities: Transformative pathways towards sustainable urban development*. United Nations Economic and Social Commission for Asia and the Pacific.

UNFCCC. (2015). *Paris agreement*. United Nations Framework Convention on Climate Change.

UNFCCC. (2025). *Guidance for NDCs 3.0 preparation*. Retrieved November 4, 2025 from https://unfccc.int/

UN-Habitat, SDU Resilience, & UNDP. (2024). *Urban content of NDCs: Local climate action explored through in-depth country analysis 2024 Report*. Global Covenant of Mayors for Climate & Energy. [Lead authors: Nicola Tollin, James Vener, Yu Liu, Patrizia Gragnani, Maria Pizzorni, and Bernhard Barth]. Retrieved November 4, 2025 from https://unhabitat.org/sites/default/files/2025/09/ndc_global_report_2024.pdf

UN-Habitat. (2024). *World cities report 2024: Pathways to urban transformation*. United Nations Human Settlements Programme.

Vogel, J., Carioni, E., Fenner, D., & Meier, F. (2024). Urban greenhouse gas emissions: Current trends and future projections. *Nature Cities, 1*(2), 89–101.

Yoffee, N., & Terrenato, N. (2015). Introduction: A history of the study of early cities. In N. Yoffee (Ed.), *The Cambridge world history* (pp. 1–44). Cambridge University Press.

PART I

Cities and Climate Change

This section follows the evolution of the Climate Change and Cities Conference series from its inaugural 2018 meeting through Innovate4Cities 2021and 2024, and into planning for 2026. The section emphasizes the conference's founding role in bridging science, policy, and urban practice, and filling the gap in evidence on the science of cities and climate change. The conference series has established the Global Research and Action Agenda on Cities and Climate ChangeScience (GRAA) as an adaptive, evolving framework over the evolution of the conferences and the numerous connected meetings between. The GRAAseeks to support global initiatives such as the UNFCCCParis Agreement, United Nations Sustainable Development Goals, and New Urban Agenda while highlighting, generating, and supporting city knowledge for global scientific assessments. Initiatives include the IPCC's Sixth and Seventh Assessment Reports, including the Special Report on Climate Change and Cities (SRCCC), due for release in 2027. The chapters in this section look at how individuals and their communities of practice and process have carried out the GRAA agenda to integrate local and regional knowledge on solutions, challenges, and opportunities regarding the impact of climate change on urban communities into global research priorities while equipping cities with essential tools, partnerships, and professional networks to accelerate climate action.

The section also analyzes the pivotal role of boundary-spanning actors and city transnational networks in fostering innovation across the science-policy-practice divide

CHAPTER 2

Cities and Climate Change Science Conference: Bridging Research and Action

Brenna Walsh, Cathy Oke, Benjamin Jance IV, Julie Greenwalt, Debra C. Roberts, Gian Carlo Delgado-Ramos, Steven Bland, and Lea Ranalder

2.1 Introduction

The genesis of this book can be traced to the 43rd IPCC Plenary session (P43) in Nairobi, Kenya in 2016, during the preliminary phases of the Intergovernmental Panel on Climate Change (IPCC)'s Sixth Assessment Report cycle (AR6). The trajectory from this initial proposal to the 2018

B. Walsh
Vancouver, Canada

C. Oke (✉)
Melbourne Centre for Cities, Faculty of Architecture Building and Planning, University of Melbourne, Melbourne, Australia
e-mail: cathy.oke@unimelb.edu.au

B. Jance IV
Global Covenant of Mayors for Climate and Energy, Brussels, Belgium
e-mail: bjance@globalcovenantofmayors.org

C. Oke et al. (eds.), *A Global Research and Action Agenda on Cities and Climate Change Science*,
https://doi.org/10.1007/978-981-95-6517-7_2

Conference in Edmonton, through the entirely virtual Innovate4Cities 2021 Conference to the 2024 Innovate4Cities Conference in Montréal, represents a substantial evolution in global approaches to urban climate action—an evolution that this volume seeks to systematically document and analyze.

This developmental trajectory reflects not only the increasing recognition of cities as critical actors in climate change mitigation and adaptation (Chapter 1), but also the growing sophistication of collaborative frameworks between academic researchers, urban practitioners, and policymakers. At P43, 32 proposals for special reports were submitted for Panel consideration, among which was a South African proposal for a Special Report on Cities. While the urban focus was deemed policy-relevant, the Panel determined that insufficient scientific knowledge existed to support a Special Report on Cities within AR6, proposing instead that such a report be incorporated into the Seventh Assessment Report cycle (AR7). To address these identified knowledge gaps, the Panel agreed to consider

J. Greenwalt
Cities Alliance, Brussels, Belgium
e-mail: jgreenwalt@citiesalliance.org

D. C. Roberts
School of Life Sciences, University of KwaZulu-Natal, Durban, South Africa

Faculty of Geo-Information Science and Earth Observation, University of Twente, Enschede, The Netherlands

D. C. Roberts
e-mail: Debra.Roberts@palebludot.co.za

G. C. Delgado-Ramos
Institute of Geography, National Autonomous University of Mexico, Mexico City, Mexico
e-mail: giandelgado@unam.mx

S. Bland
World Resources Institute, London, UK
e-mail: Steven.bland@wri.org

L. Ranalder
UN-Habitat, Nairobi, Kenya
e-mail: lea.ranalder@un.org

collaborating with academic institutions, urban practitioners, and relevant scientific bodies and agencies to organize an international scientific conference on climate change and cities early in the AR6 cycle, with the explicit objective of stimulating scientific reports and peer-reviewed publications on this subject (IPCC, 2016).

This recommendation catalyzed the formation of a coalition comprising nine organizations focused on global cities and climate: Cities Alliance, C40 Cities Leadership Group, Future Earth, ICLEI–Local Governments for Sustainability, Sustainable Development Solutions Network (SDSN), United Cities and Local Governments (UCLG), United Nations Environment Programme (UNEP), United Nations Human Settlements Programme (UN Habitat), and the World Climate Research Programme. This coalition proposed hosting what would be designated the 2018 Conference, which was co-sponsored by the IPCC. Following a competitive selection process spanning 2016–2017, Edmonton, Canada was selected as the host city, with the City of Edmonton joining the conference organizing committee (Fig. 2.1 shows the evolution timeline from 2016 through the planned 2027 IPCC AR7 Special Report on Climate Change and Cities (SRCCC). The conference's Scientific Steering Committee was assembled through an international call for members, with selection criteria emphasizing both subject matter expertise and regional representation.

2.2 Conference Evolution

2.2.1 *The 2018 Cities and Climate Change Science Conference*

The inaugural Cities and Climate Change Science conference in Edmonton was conceptualized as a strategic platform to bridge scientific expertise from academic researchers contributing to IPCC Assessment cycles with practical knowledge from practitioners and policymakers addressing globally relevant urban challenges. This integration was essential for developing future research that could encompass both comprehensive scope and contextual specificity required to address urban climate challenges effectively (Bai et al., 2018).

The conference had three specific objectives: (1) identify key research and knowledge gaps regarding cities and climate change; (2) inspire global and regional research leading to peer-reviewed publications and scientific reports; and (3) stimulate research on cities and climate change

Key Metrics Evolution:

	2018	2021	2024	2026
Participation	700		1985	
GRAA scope	80 research gaps	120+ research gaps	160 research gaps + 259 action priorites	
Geographic reach	North America/ Europe focus	→	Global with regional tracks	
Format evolution	In-person	Virtual	Hybrid	
Urban Content within NDCs	• 2017: 16% of 164 NDCs analyzed contained a strong level of urban content, 53% had a moderate level of urban content, and 31% had no urban content • 2021: 45% of 157 NDCs analyzed contained a strong level of urban content, 39% had a moderate level of urban content, and 16% had no urban content • 2023: 27% of 194 NDCs analyzed contained a strong level of urban content, 39% had a moderate level of urban content, and 34% had no urban content			

IPCC-related milestones (left-hand side of the timeline) are included solely to situate the development of the GRAA and the Innovate4Cities initiative across the broader time frame of key institutional milestones. Mentions of the IPCC do not constitute any formal linkage with the GRAA and Innovate4Cities initiative, which are independent processes informed by organic interactions that may occur among city networks, institutions, governments, and scientists engaged in multiple processes.

Fig. 2.1 Evolution timeline with multi-level impact (2016–2027) (*Source* Authors)

throughout AR6. The primary deliverable was the Global Research and Action Agenda on Cities and Climate Change Science (GRAA), articulating specific research gaps across four cross-cutting areas, six topical areas, and three delivery approaches (Prieur-Richard et al., 2018).

The conference featured intensive deliberations across thematic areas that shaped urban climate research trajectories (United Nations Human Settlements Programme, 2018). Central debates addressed scalar integration, balancing global research frameworks with contextual specificity for local action. Urban governance sessions revealed tensions between top-down policy approaches and bottom-up community initiatives, recognizing that effective urban climate action necessitates integration across multiple governance levels.

Technological discussions focused on urban infrastructure's dual function as both contributor and solution to climate challenges, emphasizing the potential of smart-city technologies for responsive and adaptive urban systems. Financial sessions highlighted the critical gap between climate ambition and available funding mechanisms, generating calls for innovative financing approaches across diverse urban contexts.

Most significantly, discussions revealed the critical importance of informal settlements and informal economy actors in urban climate responses—issues systematically underrepresented in previous climate research but essential for addressing climate challenges in rapidly growing Global South cities. Visa-issuance delays prevented some Global South participation, subsequently influencing enhanced emphasis on justice and equity in later iterations.

Quote Box 1—2018 Edmonton:

> The conference provided a benchmark that simultaneously illustrates the shift from a research agenda largely focused on projecting climate impacts on sectors and regions, to an adaptation science studying what to do about them.
> —Solecki et al. (2021)

Conference composition demonstrated that the majority of attendees came from academia: 48% academic institutions, 22% practitioners, and 13% policymakers (Hunter et al., 2022). IPCC bureau members assumed multiple leadership roles, with two Working Group Co-Chairs and a Working Group Vice-Chair on the Scientific Steering Committee, three Technical Support Unit representatives, and an IPCC Secretariat member on the Organising Committee. The Scientific Steering Committee maintained ultimate decision-making authority regarding the program, substantially shaping outcomes including the GRAA (Co-Chairs of the Scientific Steering Committee, 2018).

With fewer than 10 city mayors attending, the resulting GRAA was structured more for academic researchers. Following the conference, a member of the Scientific Steering Committee approached the Global Covenant of Mayors for Climate and Energy (GCoM) to propose bridging the gaps in the GRAA by further engaging representatives of cities and city networks. This established the Innovate4Cities initiative, producing the City Research Agenda (CRA) through six global workshops bringing together academics, mayors, and practitioners—particularly private-sector representatives, who had been underrepresented in the 2018 conference (Hunter et al., 2022).

2.2.2 *Evolution to Innovate4Cities 2021*

The Innovate4Cities initiative, launched by GCoM in 2018 (Fig. 2.2), centers on identifying data, information, and technology gaps supporting city climate action. In 2019, GCoM established the Research and Innovation Technical Working Group (TWG), co-chaired with UN-Habitat, providing targeted guidance on research gaps and collaborative frameworks for the Innovate4Cities initiative (GCoM, 2019). In 2020, the TWG started planning a follow-up conference, which end up being conducted virtually in 2021 due to COVID-19 and was again co-sponsored by the IPCC. The series was renamed as the "Innovate4Cites Conference" thereafter.

The objectives articulated in the 2021 conference evolved substantially, with the conference accommodating lessons learned and pandemic circumstances to focus on research and innovation while "closing the gap between ambition and implementation." Three objectives emerged: (1) to advance progress on GRAA gaps, particularly those related to COVID-19; (2) to catalyze scientist-practitioner engagement for UNFCCC inputs and in support of the more robust and integrated focus on urban issues in the IPCC's AR6; and (3) to generate platforms for co-creating and disseminating policy, technology, and organizational innovations for

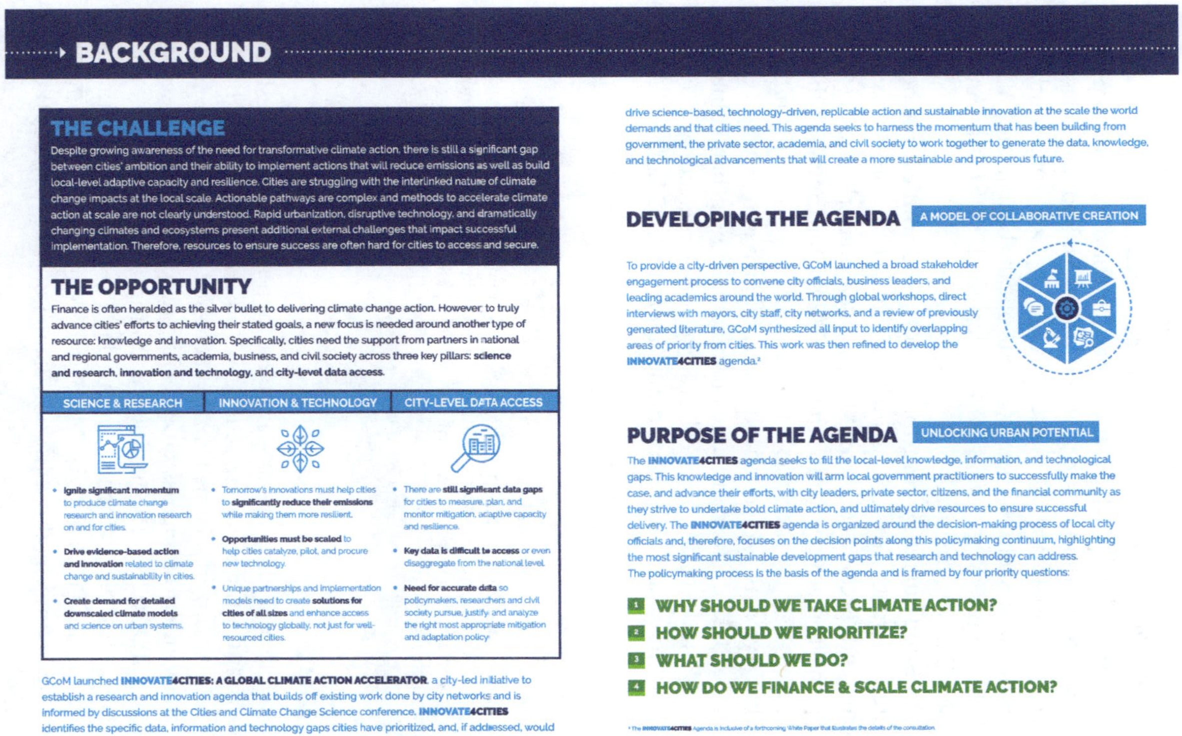

BACKGROUND

THE CHALLENGE

Despite growing awareness of the need for transformative climate action, there is still a significant gap between cities' ambition and their ability to implement actions that will reduce emissions as well as build local-level adaptive capacity and resilience. Cities are struggling with the interlinked nature of climate change impacts at the local scale. Actionable pathways are complex and methods to accelerate climate action at scale are not clearly understood. Rapid urbanization, disruptive technology, and dramatically changing climates and ecosystems present additional external challenges that impact successful implementation. Therefore, resources to ensure success are often hard for cities to access and secure.

THE OPPORTUNITY

Finance is often heralded as the silver bullet to delivering climate change action. However, to truly advance cities' efforts to achieving their stated goals, a new focus is needed around another type of resource: knowledge and innovation. Specifically, cities need the support from partners in national and regional governments, academia, business, and civil society across three key pillars: **science and research, innovation and technology**, and **city-level data access**.

SCIENCE & RESEARCH	INNOVATION & TECHNOLOGY	CITY-LEVEL DATA ACCESS
• **Ignite significant momentum** to produce climate change research and innovation research on and for cities. • **Drive evidence-based action and innovation** related to climate change and sustainability in cities. • **Create demand for detailed downscaled climate models** and science on urban systems.	• Tomorrow's innovations must help cities to **significantly reduce their emissions** while making them more resilient. • **Opportunities must be scaled** to help cities catalyze, pilot, and procure new technology. • Unique partnerships and implementation models need to create **solutions for cities of all sizes** and enhance access to technology globally, not just for well-resourced cities.	• There are **still significant data gaps** for cities to measure, plan, and monitor mitigation, adaptive capacity and resilience. • **Key data is difficult to access** or even disaggregate from the national level. • **Need for accurate data** so policymakers, researchers and civil society pursue, justify and analyze the right most appropriate mitigation and adaptation policy.

GCoM launched **INNOVATE4CITIES: A GLOBAL CLIMATE ACTION ACCELERATOR**, a city-led initiative to establish a research and innovation agenda that builds off existing work done by city networks and is informed by discussions at the Cities and Climate Change Science conference. **INNOVATE4CITIES** identifies the specific data, information and technology gaps cities have prioritized, and, if addressed, would drive science-based, technology-driven, replicable action and sustainable innovation at the scale the world demands and that cities need. This agenda seeks to harness the momentum that has been building from government, the private sector, academia, and civil society to work together to generate the data, knowledge, and technological advancements that will create a more sustainable and prosperous future.

DEVELOPING THE AGENDA — A MODEL OF COLLABORATIVE CREATION

To provide a city-driven perspective, GCoM launched a broad stakeholder engagement process to convene city officials, business leaders, and leading academics around the world. Through global workshops, direct interviews with mayors, city staff, city networks, and a review of previously generated literature, GCoM synthesized all input to identify overlapping areas of priority from cities. This work was then refined to develop the **INNOVATE4CITIES** agenda.[2]

PURPOSE OF THE AGENDA — UNLOCKING URBAN POTENTIAL

The **INNOVATE4CITIES** agenda seeks to fill the local-level knowledge, information, and technological gaps. This knowledge and innovation will arm local government practitioners to successfully make the case, and advance their efforts, with city leaders, private sector, citizens, and the financial community as they strive to undertake bold climate action, and ultimately drive resources to ensure successful delivery. The **INNOVATE4CITIES** agenda is organized around the decision-making process of local city officials and, therefore, focuses on the decision points along this policymaking continuum, highlighting the most significant sustainable development gaps that research and technology can address. The policymaking process is the basis of the agenda and is framed by four priority questions:

1. **WHY SHOULD WE TAKE CLIMATE ACTION?**
2. **HOW SHOULD WE PRIORITIZE?**
3. **WHAT SHOULD WE DO?**
4. **HOW DO WE FINANCE & SCALE CLIMATE ACTION?**

[2] The INNOVATE4CITIES Agenda is inclusive of a forthcoming White Paper that illustrates the details of the consultation.

Fig. 2.2 Excerpt from first Innovate4Cities Booklet 2018 (Global Covenant of Mayors for Climate and Energy, 2019, p2)

climate-neutral, resilient cities. The GRAA and CRA (the latter later evolving into the City Research *and Innovation* Agenda) were both updated after the 2021 conference (Oke et al., 2022; Walsh et al., 2022).

The 2021 format facilitated enhanced participation spanning 72 hours across three days to accommodate global time zones. Four regional tracks were covered: (1) Oceania and Southeast Asia; (2) East Asia, South Asia, Eastern Europe, and Central Asia; (3) European Union and Western Europe, Middle East, West Asia, and Africa; and (4) Latin America, Caribbean, and North America.

COVID-19 fundamentally transformed urban priorities, introducing unprecedented urgency to discussions of resilience, health systems, digitalization, and green recovery. "Build Back Better" and a "green and climate-smart recovery from the COVID-19 pandemic" became prevailing terms within the discourse, emphasizing opportunities to redesign urban systems for enhanced sustainability and equity (GCoM & UN-Habitat, 2022). The pandemic illuminated how urban vulnerabilities—air quality, housing density, healthcare access—intersect with climate risks, leading to integrated urban-planning approaches (Mahadevia et al., 2025).

Quote Box 2—2021 Virtual:

> The virtual format democratised participation by removing travel barriers and costs, enabling unprecedented global engagement while also changing the nature of collaboration and informal networking that had characterised in-person gatherings.
> —Conference Synthesis Report (2021)

Despite initial optimism about pandemic opportunities for reshaping work patterns and urban spaces, scholarly consensus indicates that this potential was unrealized. Transformative change has remained limited despite meaningful efforts across cities, with some experiencing setbacks (Mahadevia et al., 2025). Inequality has continued to increase, greenhouse gas emissions have reached record levels, and coal consumption has continued to rise (IEA, 2025). Major oil companies are withdrawing from renewable commitments, with only China on track for 2030 targets (REN21, 2024).

2.2.3 *Innovate4Cities Conference, Montréal: Bridging Ambition and Implementation*

Following 2021, the Research and Innovation Technical Working Group disseminated results through publications like "Findings from Innovate4Cities 2021 and Update to the Global Research and Action Agenda" (2022). Recognizing the need for sustained interdisciplinary dialogue and intensified problem owner-solution provider engagement, regional marketplaces were organized in 2023 to inform GRAA priorities.

Montréal was selected for the 2024 conference, given its leadership in integrated climate-biodiversity action. The conference aimed to accelerate urban climate action by bridging implementation gaps, shaping global discourse, and fostering collaborative solutions through empowering cities, enhancing cross-sectoral partnerships, and benefitting from regional voices. This was the first conference in the series not co-sponsored by the IPCC, marking a move away from the strong scientific focus of the previous two events.

The 2024 hybrid model accommodated 752 in-person participants and 1233 virtual attendees, and featured 217 abstracts. This balanced the advantages of in-person collaboration with global accessibility, though somewhat reducing 2021's regional focus. A growth in participation from approximately 700 in 2018 to nearly 2000 in 2024 reflects both an expanding recognition of cities as critical climate actors and the initiative's success in building a global community of practice (Table 2.1 gives the participant breakdown through all three conferences, although numbers may not add to total participants as some may identify as members of multiple groups).

Quote Box 3—2024 Montréal:

> The building structure innovation and unified GRAA achievement represents a maturation from linear thinking to complex, systems-based approaches that better capture the multifaceted nature of urban climate challenges.
> —2024 Conference Synthesis

Table 2.1 Summary of participant and demographic numbers by sector (% as a proportion of total participants per year)

	2018	*2021*	*2024*
Academia	323 (46.1%)	2271 (32.9%)	490 (24.7%)
Business	3 (0.4%)	1767 (25.6%)	284 (14.3%)
Government	141 (20.1%)	1187 (17.2%)	354 (17.8%)
Civil Society	148 (21.1%)	179 (2.6%)	492 (24.8%)
International Organization		352 (5.1%)	303 (15.3%
Non-Governmental Organization		1056 (15.3%)	303 (15.3%)
Media	39 (5.6%)	35 (0.5%)	3 (0.2%)
TOTAL	**701**	**6901**	**1985**

2.3 The Conference Series Across Literature, Discourse, and Practice on Cities and Climate Change

2.3.1 Cross-disciplinary Knowledge Generation and Sharing

The Innovate4Cities conference series is a pioneering—and constantly evolving—experiment for novel collaboration and partnerships aimed to advance and accelerate urban climate action. The conferences have been explicitly designed to break down traditional divisions between academic research, policy development, and on-the-ground practice that organizers believed had historically limited the effectiveness of climate responses in urban contexts. Built on a base of knowledge co-generation that evolves into practitioner adoption, research influence, and—where most relevant and applicable—global policy, the Innovate4Cities process helps ensure that a robust knowledge base and community of practice accompany knowledge as it evolves (Fig. 2.3). This cross-disciplinary, multi-expert, and knowledge-diverse approach operates at multiple levels: methodologically, by integrating different types of knowledge systems and evidence; institutionally, by bringing together organizations with varying mandates and approaches; and geographically, by connecting global frameworks with local implementation contexts, linking to the boundary-spanning concept further explored in Chapter 4.

The conference series has occupied a distinctive position within the broader landscape of urban climate research and practice. Hunter et al. (2022) identified four key factors that provided the temporal context

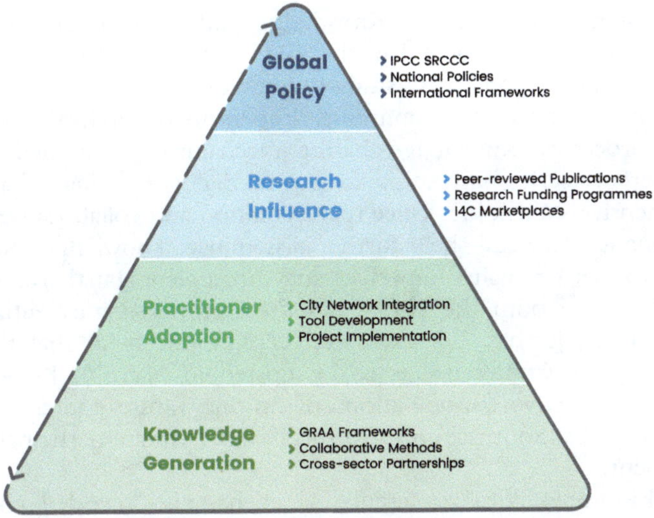

Fig. 2.3 Stakeholder ecosystem evolution and knowledge impact, from 2018–2024 (*Source* Authors)

for the proposal and success of the 2018 conference: (1) increasing recognition of urban importance in climate change discourse in 2015, including the incorporation of a cities-focused Sustainable Development Goal and enhanced emphasis on addressing climate change through the Paris Agreement; (2) institutional and funding support from conference organizing partners; (3) increased IPCC interest in cities and climate change; and (4) the host city's capacity to provide substantial funding for conference execution.

The conferences have undertaken systematic efforts to integrate Indigenous Knowledge and Local Knowledge systems with conventional scientific methodologies, recognizing that effective urban climate action necessitates these diverse knowledge systems and practices. This effort complements ongoing initiatives in multiple countries, including the host country of Canada (2018 and 2024), within the broader Innovate4Cities initiative and in IPCC processes to understand how Indigenous Knowledge systems can be systematically integrated with local and scientific knowledge frameworks.

This integration of Local Knowledge and Indigenous Knowledge was facilitated and broadened in the 2021 conference through regional tracks, community-based case study presentations, and explicit efforts to include practitioners and community leaders in research agenda development processes. Knowledge-sharing mechanisms developed through the conference series—including structured dialogue formats that ensure comprehensive stakeholder voice representation and collaborative agenda development processes—help further disseminate knowledge. Nevertheless, this process remains imperfect and incomplete. Further efforts are required, within both the IPCC and the Innovate4Cities initiative, to advance this trajectory. For example, the announcement that the 2026 Innovate4Cities Conference is to be hosted in Nairobi, Kenya marks a response to active considerations of hosting future conference iterations in a Global South country to specifically facilitate participation and involvement.

The Innovate4Cities conference series has also revealed challenges and opportunities associated with bridging different communities of practice. Solecki et al. (2021) identified distinct priorities among stakeholder groups: academics demonstrated greater interest in developing enhanced understanding of low-probability, high-consequence scenarios for key climate impacts such as sea-level rise; practitioners focused more acutely on actions with the highest probability and greatest certainty of affecting their specific cities; and urban policymakers demonstrated increased interest in understanding what other cities have attempted and the success or failure of these interventions.

Analysis by Hunter et al. (2023) found mixed success in achieving the 2018 conference goals and objectives. While practitioners utilized the first iteration of the GRAA to shape projects and programs, the GRAA was also perceived as too general to effectively meet cities' specific needs. Additionally, while new partnerships emerged between the three stakeholder groups, they did not always develop through expected mechanisms, such as the Edmonton Declaration, which had been signed at the conference (GCoM, 2018).

While a more balanced distribution of participants across research, practitioner, and policymaking communities has provided benefits in broadening conversations, partnerships, and solutions more precisely addressing specific urban needs, there has been a decrease in conference discussions and subsequent GRAA-influenced direct scientific output and publications from the 2018 iteration to 2024. Refocusing on research through collaborations, data platforms, and publications originating at the conference will help support and strengthen the scientific evidence available for the SRCCC report, and demonstrate a strong focus on a scientific evidence base to support implementation and financing.

2.3.2 *Challenges in Global Participation and Representation*

Analysis of regional representation across the conference series reveals persistent challenges in achieving genuinely global participation. The 2018 conference demonstrated a predominance of North American and European participants, reflecting traditional patterns in international academic conferences. The virtual format of the 2021 conference significantly improved Global South participation, with regional tracks providing platforms for voices that had been systematically underrepresented in 2018. However, digital divides and time zone challenges meant that participation remained uneven. The hybrid model used in 2024, while improving in-person networking opportunities, resulted in some regression in Global South representation.

This pattern highlights the persistent tension between accessibility and depth of engagement that characterizes international collaborative efforts. Solutions under exploration include dedicated funding for Global South participation, regional preparatory meetings, and more-sophisticated hybrid technologies that better integrate virtual and in-person participants, including a dynamic GRAA online platform to provide a digital space where practitioners worldwide can continually exchange information, fill knowledge gaps, and exchange feedback. As of November 2025, the GCoM Secretariat, in collaboration with the Research and Innovation Technical Working Group, is developing a City Climate Knowledge Center (working title), which, when combined with the existing GCoM Resource Library, can offer these features for a global community of researchers, practitioners, and policymakers on cities and climate.

2.3.3 *Bridging Science, Practice, and Innovation for Accelerated Action*

The conference series has been particularly innovative in developing mechanisms that explicitly bridge the science-practice-innovation nexus. Rather than treating these as separate domains, the conferences have created integrated approaches that recognize the interdependence of research, implementation, and technological and social innovation in addressing urban climate challenges (although the emphasis on the different elements of the integrated landscape has changed at each event). The science–practice bridge operates through the collaborative development of research agendas that ensure that research priorities reflect real-world implementation needs while maintaining scientific rigor. Practice–innovation connections are fostered through showcase sessions that highlight successful local innovations and explore their potential for scaling or adaptation in other contexts. The science–innovation nexus is supported through dedicated sessions on emerging technologies and their research implications, ensuring that technological development is informed by robust evidence and scientific understanding.

This integrated approach has been particularly valuable for accelerating action because it reduces the traditional temporal lag between research findings and practical implementation. By involving practitioners in research design and scientists in implementation assessment, the conferences have created more-responsive and actionable knowledge-production processes.

2.4 Development of the Global Research Action Agenda on Cities and Climate Change Science (GRAA).

The GRAA will be discussed in more detail in Chapter 3; presented here are the origins and evolution of the knowledge and action priorities, topics, and themes, as illustrated in the research agenda.

2.4.1 Original Methodology

The original 2018 GRAA was compiled from diverse sources through a structured process. Each session convener in the 2018 conference received questions to identify research gaps from their presentation and subsequent discussions, supported by session rapporteurs. Poster presenters completed similar questionnaires. The Scientific Steering Committee and Organising Committee met daily during the conference to discuss emerging themes, with GRAA authors compiling information each evening.

In the two days following the 2018 conference, the Scientific Steering Committee and Organising Committee developed the GRAA outline from conference discussions and initial data collection. A comprehensive data table compiled information from each session and poster, ensuring that authors could refer to specific research, knowledge, and action gaps identified during sessions. An initial draft of the GRAA was shared with conference participants for feedback. This process was adapted for the virtual conference in 2021, with two co-editors leading the writing of the expanded GRAA in consultation with the Scientific Steering Committee. In 2024, data was collected from conference presenters, with reactions to material collected in real time from both in-person and virtual participants.

2.4.2 Structural Evolution

In addition to changes in methods used to collect data across the three conferences, the GRAA has undergone significant structural transformation, reflecting the growing complexity and sophistication of understanding urban climate challenges. The 2018 version organized research gaps under four cross-cutting areas, six topical areas, and three delivery approaches, visualized through a concentric wheel structure.

The 2021 iteration expanded this framework by adding three additional cross-cutting areas—health, digitalization, and justice and equity—informed by Hadfield et al.'s (2021) work and mapping of progress on 2018 GRAA research gaps. The intersections of climate change and health, particularly in dense urban areas, had been emerging for some time but were highlighted by the COVID-19 pandemic. The area of history and cultural heritage was added as the only additional topical area during the 2021 conference, surfacing through regional tracks that emphasized the need for place-based action rooted in traditional practices.

The 2021 GRAA also introduced a triple-helix structure to supplement the wheel, highlighting expanded conversation and thinking about addressing gaps. This brought emphasis to needed mindset shifts: (1) envisioning the future, (2) applying an integrated systems approach, and (3) adopting people-centered solutions, alongside tools and processes including participation, partnership, finance, capacity-building, and research. Providing a basis for the building model introduced in 2024, this structure illustrated recognition that the wheel framework could not capture the full complexity of urban climate research, action, and innovation needs.

2.4.3 The 2024 Transformation

The most significant structural evolution occurred in 2024 with the introduction of a building as a spatial mnemonic device, evolving from the helix and replacing the wheel structure. This new framework comprises four key components:

1. **Delivery approaches** as methods to move between floors (staircases, lifts).
2. **Pillars of justice and equity** as the building foundation.
3. **Systems-level approaches** as floors of the building.
4. **City-level models, data, and knowledge** as information to be communicated across different floors.

This structure allows for future research and experience in addressing climate change in cities to populate the various areas of consideration with new knowledge, providing pathways for interlinkages between topics. The building structure was "stress tested" during the conference, with participants considering how newly identified knowledge gaps would fit into the framework.

A significant methodological advancement in the 2024 GRAA was the integration of previously separate agendas. Following recommendations from the Global Marketplace workshop synthesis report (Jance et al., 2024), the City Research and Innovation Agenda (CRIA) and the GRAA itself were unified into a single framework. This unified agenda (detailed in Chapter 3) marks the growth and evolution of the GRAA to better represent the full complexity of addressing climate change in cities, and provides a more useful tool for academics, practitioners, policymakers, and other stakeholders, as it focuses on both research and action.

The Innovate4Cities discourse increasingly incorporated digital collaboration tools, with the 2024 conference using Miro boards and other virtual collaboration platforms to enable real-time co-creation of the research agenda. This represented a significant evolution from the static, paper-based data-collection methods used in 2018 to digital tools that allowed for more dynamic and iterative agenda development. The hybrid conference format enabled participants to contribute to shared digital workspaces both during and between sessions, creating more opportunities for sustained collaboration beyond the formal conference program.

The quantitative growth of the GRAA reflects the expanding scope and sophistication of urban climate research. The 2018 version identified approximately 80 research gaps across its framework, which expanded to over 120 in 2021 with the addition of regional perspectives and new cross-cutting themes. The 2024 integration of research gaps and action priorities into a unified framework of 160 research gaps and 259 action priorities represents not just numerical growth but a qualitative shift toward more actionable and implementation-focused guidance. The 2024 GRAA also introduced more-sophisticated tagging systems, with research gaps and action priorities tagged with multiple topical areas rather than fitting under a single topic. This evolution reflects the field's maturation from primarily identifying what is not known to providing clearer pathways for what needs to be done, through the inclusive feedback system the Innovate4Cities initiative has encouraged.

2.4.4 *Alignment with the IPCC Special Report on Climate Change and Cities (SRCCC)*

The conference series has played a crucial role in building the scientific foundation for the SRCCC, demonstrating how sustained collaborative effort can influence global scientific assessment processes. The GRAA frameworks and conference-outcome reports have contributed towards informing the structure and scope of the SRCCC, with many of the research gaps identified through the conference process now being assessed by the 97 expert authors from 56 countries selected to develop the report, five of whom are city practitioners—a significant advancement.

The conference series has also contributed to capacity-building within the global urban climate research community, with many conference participants going on to serve as authors, reviewers, or contributors to the IPCC process. This demonstrates how collaborative platforms can serve not only to identify research needs, but also to develop the necessary human capital to ensure that participants are considered in assessing progress. The regional focus introduced in 2021 highlighted the importance of reflecting diverse urban contexts and challenges within the SRCCC rather than being dominated by experiences from a limited number of cities or regions.

2.4.5 *The Impact of the Riga Scoping Meeting*

The initial scoping meeting for the SRCCC was held in Riga, Latvia in April 2024 (IPCC 2024). IPCC leadership made a deliberate decision to invite urban practitioners. Key impacts of the Innovate4Cities initiative on the IPCC process include:

- **Coordinated inputs:** The existing networks enabled more structured and coordinated inputs into the scoping process.
- **Reality-check function:** The practitioner voice often served as a reality check against some of the more theoretical aspirations of academia
- **Solutions focus influence:** The practitioner voice was instrumental in the emergence of Chapter 5: Solutions by City Type and Region.

2.5 Where Do We Go From Here?

A central objective of the conference series has been to provide cities and local governments with institutional spaces, professional networks, strategic partnerships, and analytical tools necessary to accelerate climate ambition and implementation. This objective is being achieved through multiple mechanisms operating at different scales and over different timeframes.

The conferences themselves function as networking spaces where city practitioners can establish connections with researchers working on relevant topics, leading to collaborative projects and knowledge-exchange relationships that extend substantially beyond the conference events. The GRAA frameworks provide cities with structured approaches to identifying research needs and connections with academic partners who can assist in addressing them. The regional tracks introduced in 2021 have been particularly valuable for facilitating connections between cities facing similar challenges within specific geographic or cultural contexts.

The Innovate4Cities initiative has created ongoing platforms for sustained collaboration, including the Research and Innovation Technical Working Group, which provides a permanent mechanism for cities to engage with researchers and innovation partners. The Marketplace workshops that preceded the 2024 conference exemplify this approach, bringing together cities, researchers, and innovation providers around specific thematic areas to develop collaborative projects and partnerships.

2.5.1 Impact on Science, Policy, and Practice

Evidence of the initiative's impact can be documented across multiple domains.

In academia, there is a growing emergence of peer-reviewed publications that apply the GRAA and its themes and collaboration mechanisms in distinct, yet complementary, disciplines (Daudin et al., 2021; Liu & Masago, 2023; Moreira et al., 2022). Research and innovation funding programs are also integrating GRAA-aligned themes and topics into their priorities and criteria, highlighting an appetite for evidence-driven implementation opportunities; an example is the Driving Urban Transitions Partnership (JPI Urban Europe, 2022).

In policy, joint statements emerging from Innovate4Cities conferences—including the Edmonton Declaration in 2018 and the Montréal Call to Action in 2024—exemplify the increasing precedent for science-informed advocacy and policy influence (GCoM, 2018; UN-Habitat, 2024). The latter's inclusion of ties to the Coalition for High Ambition Multilevel Partnerships (CHAMP)—a platform to enhance collaboration between national and subnational governments in climate policy and finance launched at COP28 by the COP28 Presidency and Bloomberg Philanthropies—suggests stronger opportunities for engagement and alignment with multi-level governance and international frameworks (discussed further in Chapter 11).

Across practice, financing opportunities and new resources are building on priorities built into the GRAA. Flagship examples include the City Climate Finance Gap Fund and the C40 Cities Finance Facility, both of which provide targeted technical and capacity-focused support to local governments at varying stages of the project-development pipeline (Gogoi et al., 2018; Irvin et al., 2024). Looking ahead, the "building" analogy of the 2024 GRAA is intended to help co-generate and capture knowledge and resources that can support local practitioners, with an aim of "populating" its floors and spaces through the Innovate4Cities process (Irvin et al., 2024).

2.5.2 Ongoing Challenges and Future Directions

The unified GRAA emerging from the 2024 Innovate4Cities conference provides a more comprehensive framework for addressing urban climate challenges. Its building structure offers flexibility for future development while maintaining coherence across the complex landscape of urban climate research and action. The framework's emphasis on justice and equity as foundational pillars reflects growing recognition that effective climate action must address fundamental questions of fairness and inclusion for systems to function sustainably.

However, significant challenges, such as the tension between the general nature of global frameworks and the specific needs of individual cities, continue to require attention. The need for greater Global South representation and voice remains an ongoing concern.

Additionally, the challenge of translating research into actionable policies and programs that can be implemented at the scale and speed required by the climate crisis continues to be difficult to balance.

Crucial to the initiative's future is the sufficient, adequate, and relevant engagement of science, especially as the world approaches climate overshoot. Strengthening the scientific foundation again represents one of the most substantial challenges—and opportunities—for impact.

2.5.3 Future Opportunities

As the initiative advances, the lessons learned from this six-year evolution provide valuable insights for similar efforts to bridge science and practice to respond to complex global challenges (Irvin et al., 2024). These include:

- The success of virtual and hybrid conference formats, which has opened new possibilities for inclusive participation while building on in-person collaboration;
- Balancing accessibility with depth of engagement, including actionable takeaways for scientists, policymakers, and practitioners alike;

- Exploring and strengthening solutions concerning dedicated funding for Global South participation, regional preparatory meetings, and digital repositories for collaboration; and
- Improved communication of conference outcomes beyond practitioners to the broader public, with the aim of supporting community-scale understanding of knowledge gaps and action priorities.

The emphasis on iterative development, inclusive participation, and adaptive frameworks offers a replicable model for other domains seeking to mobilize collective action to meet shared challenges.

2.6 Conclusion

The trajectory from 2018 to 2024 has demonstrated that sustained, collaborative effort can construct substantive bridges between communities that have traditionally operated in isolation. As cities continue to confront the dual challenges of climate mitigation and adaptation, the institutional foundations established through this process provide a robust platform for the continued evolution of urban climate science and practice. The transformation from a singular, research-focused conference to a comprehensive innovation platform reflects the broader maturation of the urban climate field—progressing from identifying research gaps to actively bridging implementation priorities and demonstrating the potential for collaboration between science, policy, and practice.

The evolution of the GRAA from a wheel structure to a building metaphor symbolizes this advancement from linear thinking to more complex, systems-based approaches that can capture the multifaceted nature of urban climate challenges, and acknowledging interrelation of outcomes. Perhaps most significantly, this process has demonstrated the efficacy of sustained, iterative collaboration. The willingness of diverse stakeholders to repeatedly convene, reassess, and adapt their approaches over multiple conference cycles has created a learning ecosystem that extends substantially beyond traditional academic or policy processes.

The challenges encountered throughout this evolution—from ensuring adequate Global South representation to balancing global frameworks with local specificity—have themselves become valuable learning experiences that inform broader efforts to strengthen global environmental governance. The persistent tensions between academic rigor and practical applicability, between global coherence and regional specificity, and between ambitious vision and implementable action continues to generate productive dialogue that advances the field.

The institutional foundations established through this process position the urban climate community to make meaningful contributions to the IPCC's SRCCC and to support the growing recognition of cities as critical actors in global climate governance. The unified GRAA provides a robust framework for continued development, while growing networks and partnerships offer mechanisms for sustained collaboration.

The success of this initiative offers valuable lessons for other domains seeking to bridge the gap between science and practice in meeting complex global challenges. The emphasis on inclusive participation, iterative development, and adaptive frameworks provides a replicable model for collaborative knowledge production and action planning. As the world confronts increasing environmental and social challenges that require coordinated responses across traditional boundaries, the evolution from the Cities and Climate Change Science Conference to the Innovate4Cities conference series offers both inspiration and practical guidance for building the collaborative capacity necessary to address twenty-first-century challenges.

This journey demonstrates that meaningful progress on complex global challenges is achievable when diverse communities make a commitment to sustained collaboration, adaptive learning, and shared purpose. The institutional foundations established through this process will continue to support urban climate action for years to come while also offering a model to address other pressing challenges through similar collaborative approaches. The path forward requires continued commitment to the principles of collaboration, inclusion, and adaptive learning that have characterized this remarkable journey from a simple proposal in an IPCC plenary session in Nairobi to a global movement for urban climate action.

Acknowledgements The authors would like to acknowledge the contributions of Seth Schultz to the early formation and development of the Innovate4Cities initiative and conference series. The authors thank Veronica Castiglione for assistance with this chapter.

Competing Interests
The authors have no conflicts of interest to declare that are relevant to the content of this chapter.

References

Bai, X., Dawson, R. J., Ürge-Vorsatz, D., Delgado-Ramos, G. C., Barau, A. S., Dhakal, S., Dodman, D., Leonardsen, L., Masson-Delmotte, V., Roberts, D., & Schultz, S. (2018). Six research priorities for cities and climate change. *Nature Commentary, 23–25*, 555. https://doi.org/10.1038/d41586-018-02409-z

Co-Chairs of the Scientific Steering Committee. (2018). *Progress report. International Conference on Climate Change and Cities*. Forty-eighth session of the IPCC, Incheon, Republic of Korea, 1–5 October 2018. https://www.ipcc.ch/site/assets/uploads/2018/12/100920181041-INF1Rev1CitiesReport.pdf. Retrived on 17February2026.

Daudin, K., Weber, C., Colin, F., Cernesson, F., Maurel, P., & Derolez, V. (2021). The collaborative process in environmental projects, a place-based coevolution perspective. *Sustainability, 13*(15), 8526. https://doi.org/10.3390/su13158526

Global Covenant of Mayors for Climate and Energy, & UN-Habitat. (2022). *Discussions and findings from the 2021 Innovate4Cities conference: New insights for the 2018 global research and action agenda on cities and climate change science*. https://doi.org/10.26188/62795d16f3bab

Global Covenant of Mayors for Climate and Energy. (2018). *Edmonton declaration: Innovate4Cities—A global climate action accelerator*. https://www.globalcovenantofmayors.org/wp-content/uploads/2018/05/Edmonton_Declaration_Update_May23_v2.pdf. Retrived on 17February2026.

Global Covenant of Mayors for Climate and Energy. (2019). *Innovate4Cities: A global climate action accelerator: Research and innovation priorities*. https://www.globalcovenantofmayors.org/wp-content/uploads/2019/02/GCoM_Innovate4Cities-OPS_Booklet_8.5x11.pdf. Retrived on 17February2026.

Gogoi, E., Bisht, H., Das Roy, R., & Sharma, D. (2018). Transforming cities and finance to address climate change. In *Oxford policy management on behalf of the C40 cities finance facility (CFF)*. https://c40cff.org/knowledge-library/transforming-cities-and-finance-to-address-climate-change. Retrived on 17February2026.

Hadfield, P., Oke, C., & Verbeeck, J. (2021). *Regional research and innovation for city climate action: Global synthesis report*. Connected Cities Lab, the University of Melbourne. https://doi.org/10.26188/14743125

Hunter, N., Roberts, D., Sutherland, C., & Slotow, R. (2022). Co-creation between cities and climate change science achieves research and action agenda. *Current Research in Environmental Sustainability, 4*, Article 100189. https://doi.org/10.1016/j.crsust.2022.100189

Hunter, N., Walsh, B., Delgado Ramos, G., Greenwalt, J., Roberts, D., Oke, C., Nielson, D., & Stolow, R. (2023). Cities and climate change science after Edmonton: Knowledge, people, and organisations' influence and outreach. *Current Research in Environmental Sustainability, 6*, Article 100231. https://doi.org/10.1016/j.crsust.2023.100231

IEA. (2025). *Global energy review 2025*. https://iea.blob.core.windows.net/assets/5b169aa1-bc88-4c96-b828-aaa50406ba80/GlobalEnergyReview2025.pdf. Retrived on 17February2026.

IPCC. (2016). *Report of the forty-third session of the IPCC*. Nairobi, Kenya, 11–13 April 2026. https://www.ipcc.ch/site/assets/uploads/2018/05/final_report_p43.pdf. Retrived on 17February2026.

IPCC. (2024). *Decisions adopted by the panel (decision IPCC-LXI-5). 61st session*. https://www.ipcc.ch/site/assets/uploads/2024/08/IPCC-61_decisions-adopted-by-the-Panel.pdf. Retrived on 17February2026.

Irvin, A., Oke, C., Khan, S., Jance IV, B., Ranalder, L., Barth, B., Deacon, A., Abusamra, R., Addison, A., Adebisi, J., Adsett, C., Ailio, E., Al Jadaa, D., Aslan, S., Assalini, S., Assarkhaniki, Z., Astolfo, J. P., Badino, M., Bai, X., …, & Williams, A. (2024). *Next-generation city climate action: Findings from the 2024 Innovate4Cities conference and update to the global research and action agenda on cities and climate change science (GRAA)*. Global Covenant of Mayors for Climate and Energy, UN-Habitat & University of Melbourne. https://doi.org/10.26188/27147612

Jance, B., Irvin, A., Oke, C., Assalini, S., Barth, B., Deacon, A., Jones, A., Jones-Langley, J., Khan, S., Mariani, P., Melica, G., Oussidhoum, F., Ranalder, L., Parry, L., Salehi P., & Tacconi, M. (2024). *Shared spaces for urban innovation: Insights and outcomes from Innovate4Cities marketplaces*. Global Covenant of Mayors for Climate & Energy. https://www.globalcovenantofmayors.org/wp-content/uploads/2024/02/I4C-Marketplaces-Outcomes-Insights-Report_.pdf. Retrived on 17February2026.

JPI Urban Europe. (2022). *Driving urban transitions—Sustainable future for cities*. https://jpi-urbaneurope.eu/driving-urban-transitions-to-a-sustainable-future-dut/. Retrived on 17February2026.

Liu, F., & Masago, Y. (2023). An analysis of the spatial heterogeneity of future climate change impacts in support of cross-sectoral adaptation strategies in Japan. *Climate Risk Management, 41*, Article 100528. https://doi.org/10.1016/j.crm.2023.100528

Mahadevia, D., Delgado Ramos, G. C., Barnes, J., Fitzgerald, J., Kamei, M., & Lanza, K. (2025). *Learning from COVID-19 for climate-ready urban transformations. Elements in climate change and cities*. Cambridge University Press. https://www.cambridge.org/core/elements/learning-from-covid19-for-climateready-urban-transformation/3B01330DCDDA163C271ECE4A28AC68DB. Retrived on 17February2026.

Moreira, F. A. de A., Dalla Fontana, M., Sepe, P. M., Lopes, M. V., do Vale Moura, L., Santos Medeiros, L., de Kraker, J., Malheiros, T. F., & Di Giulio, G. M. (2022). Co-creating sustainability indicators for the local water–energy–food nexus. *Sustainability Science*, *17*(6), 2315–2329. https://doi.org/10.1007/s11625-022-01141-y

Oke, C., Walsh, B., Assarkhaniki, Z., Jance, B., Deacon, A., & Lundberg, K. (2022). The city research and innovation agenda: Prioritizing knowledge gaps and policy processes to accelerate city climate action. *Journal of City Climate Policy and Economy, 1*(1), 94–110. https://doi.org/10.3138/jccpe-2022.1.1.0007

Prieur-Richard, A.-H., Walsh, B., Craig, M., Melamed, M. L., Colbert, M., Pathak, M., Connolly, D., Leonardsen, D., Wright, G., Oke, C., Parnell, J., Okem, A., Hayward, B., Rosenzweig, C., Grimm, N. B., Moss, R., Seto, K. C., & Solecki, W. (2018). *Global research and action agenda on cities and climate change science—Full version*. https://www.ipcc.ch/site/assets/uploads/2018/12/GRAA-Cities-and-Climate-Change-Science-Full-Report.pdf. Retrived on 17February2026.

Ren21. (2024). *Renewables 2024 global status report*. https://www.ren21.net/gsr-2024/modules/global_overview/02_policy/#top. Retrived on 17February2026.

Solecki, W., Griffith, C., Larsen, L., Revi, A., Romero-Lankao, P., Smit, W., Fragkias, L., Anguelovski, E., Arslan, U., Barbieri, R., Bazán, L. M., Cobbinah, C., De Sherbinin, A., Elmqvist, P., Lwasa, S., Mende, B., Mulugetta, Y., Naess, K., Nilsson, M., ... Kernaghan, A. (2021). Growing exclusive knowledge inclusive of cities: From historical impacts to future pathways. *Current Research in Environmental Sustainability, 3*, Article 100038. https://doi.org/10.1016/j.crsust.2021.100038

UN-Habitat. (2024). *2024 Innovate4Cities conference ends with call to action on cities and climate*. https://unhabitat.org/news/13-sep-2024/2024-innovate4cities-conference-ends-with-call-to-action-on-cities-and-climate. Retrived on 17February2026.

United Nations Human Settlements Programme. (2018). Proceedings document, international conference on climate change and cities. In J. Mavoungu & M. Meyers (Eds.), *Online resource united nations human settlements programme*. https://unhabitat.org/international-conference-on-climate-change-and-cities-proceedings-document-0. Retrived on 17February2026.

Walsh, B., Greenwalt, J., Oke, C., Hadfield, P., Dickey, A., H Craig, M., North, M., Hunter, N., Okem, A., Bland, S., Verbeeck, J., Barth, B., Deacon, A., Roberts, H., Randhawa, J., Joshi, S., Silva, D. C., Lootens, M., Adler, M., &... Zhu, S. (2022). *Findings from Innovate4Cities 2021 and update to the global research and action agenda (Version 3)*. The University of Melbourne. https://doi.org/10.26188/19624029.v3

CHAPTER 3

The Global Research and Action Agenda on Cities and Climate Change Science (GRAA)

Andrew Irvin, Julie Greenwalt, Benjamin Jance IV, Cathy Oke, Samia Khan, Zahra Assarkhaniki, Brenna Walsh, Bernhard Barth, Pourya Salehi, and Jorn Verbeeck

3.1 Introduction

The Global Research and Action Agenda on Cities and Climate Science (GRAA)—the central component of the Innovate4Cities initiative—reflects the proceedings of the Innovate4Cities conferences, structured

A. Irvin (✉) · C. Oke
Melbourne Centre for Cities, Faculty of Architecture Building and Planning, University of Melbourne, Melbourne, Australia
e-mail: irvin.a@unimelb.edu.au

C. Oke
e-mail: cathy.oke@unimelb.edu.au

J. Greenwalt
Cities Alliance, Brussels, Belgium

C. Oke et al. (eds.), *A Global Research and Action Agenda on Cities and Climate Change Science*,
https://doi.org/10.1007/978-981-95-6517-7_3

through relationships between topics of central concern to the achievement of climate-resilient development in the global range of urban environments.

In relation to the broader Innovate4Cities initiative, the GRAA builds upon reflections regarding the 2018 Cities and Climate Change Science Conference and the subsequent 2021 and 2024 Innovate4Cities conferences, as presented in Chapter 2. The GRAA exists as a guidance framework for discussing "steps in a longer journey to explore the opportunities offered by evidence-based knowledge in helping address challenges associated with climate change in urban areas" (Prieur-Richard et al., 2019, p. 2) of relevance to the mission of Innovate4Cities.

Although the GRAA is complementary to some parallel systems of urban/climate knowledge acquisition and analysis (Montfort et al., 2025), it must be positioned differently in relation to other multilateral

---

e-mail: jgreenwalt@citiesalliance.org

B. Jance IV · S. Khan
Global Covenant of Mayors for Climate and Energy, Brussels, Belgium
e-mail: bjance@globalcovenantofmayors.org

S. Khan
e-mail: skhan@gcomprojectsupport.org

Z. Assarkhaniki
Faculty of Spatial Sciences, University of Groningen, Groningen, Netherlands
e-mail: z.assarkhaniki@rug.nl

B. Walsh
Vancouver, Canada
e-mail: brenna.walsh42@gmail.com

B. Barth
UN-Habitat, Nairobi, Kenya
e-mail: bernhard.barth@un.org

P. Salehi
ICLEI - Local Governments for Sustainability, Bonn, Germany
e-mail: pourya.salehi@iclei.org

J. Verbeeck
Dark Matter Labs, Amsterdam, The Netherlands
e-mail: jorn@darkmatterlabs.org

frameworks. The GRAA was introduced to implement multilateral initiatives such as the Paris Agreement, the 2030 Sustainable Development Agenda, the New Urban Agenda, and the Sendai Framework for Disaster Risk Reduction at a local level, with a framework that, "recognised the key role of cities in the global response to climate change and proposed that the [IPCC's Seventh Assessment Report (AR7)] cycle include a Special Report on Climate Change and Cities" (SRCCC) (Prieur-Richard et al., 2019, p. 1). This recognition was driven in no small part by the combined efforts of the Scientific Steering Committees for the Conferences and GCoM's Research and Innovation Technical Working Group detailed in Chapter 2.

The GRAA has emerged as a living, iterative structure, undergoing revisions and incorporating insights from the urban community via Innovate4Cities. When the GRAA was formulated initially, the exercise took place in response to "a need to develop new ways for cities and climate change science to work together with innovative, flexible and iterative processes to develop and implement solutions at the local level" (Prieur-Richard et al., 2019, p. 7). The capacity for conceptual inclusivity and consideration of additional dimensions of complexity has been reinforced with each examination of the GRAA and its ability to encompass the full range of climate change drivers faced in urban-governance scenarios when dealing with the complex interaction between sectoral actors with a range of interrelated priorities (Fig. 3.1).

3.2 Setting the Stage for Urban Research and Action: The 2018 GRAA

3.2.1 Background and Context

Prieur-Richard et al. (2019, p. 1) write, "The aim of the [Cities and Client Change Science Conference] was to assess the current state of academic, policy and practice-based knowledge on cities and climate change, and to identify key gaps to inspire research and the development of knowledge in critical areas." The conference was organized and delivered by two central committees: the Scientific Steering Committee and the Organising Committee.

The Scientific Steering Committee had the primary responsibility of shaping the conference structure: it had sole discretion in selecting conference sessions, pulling together the plenary sessions, and selecting

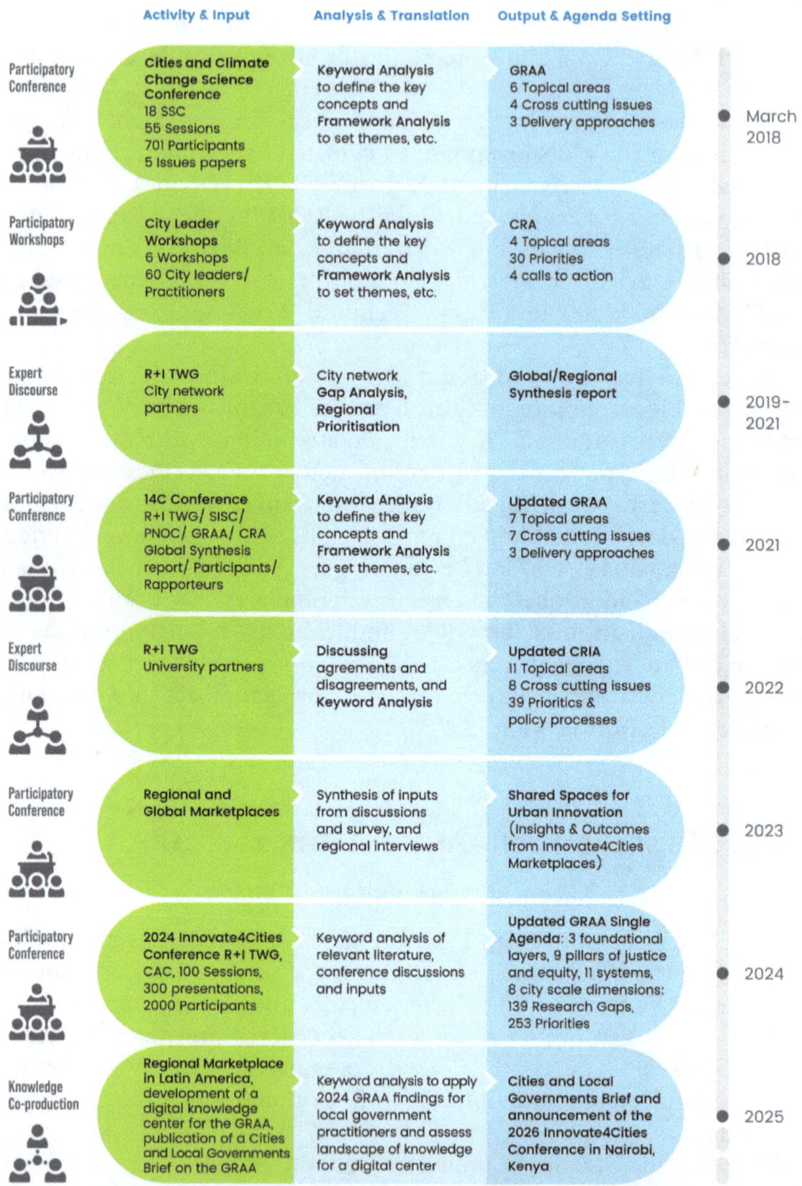

Fig. 3.1 Iterative review process for the development of the GRAA and City Research and Innovation Agenda (CRIA), timeline updated to include 2025 (*Source* Irvin et al., 2024, p. 39, used with permission)

which posters would be presented. Around 78% of the Scientific Steering Committee were academics (Hunter et al., 2022), and many had roles in the Sixth Assessment Report cycle (AR6), including two Co-Chairs and one Working Group Vice-Chair.

As the conference program began to take shape, the Scientific Steering Committee members raised some concern about gaps in the literature that could result in key topics—from the perspective of these expert committee members—not being well represented in the conference program. The committee decided together on a set of topics on which they would like to solicit commissioned discussion papers, which would prepare conference attendees for a more well-rounded conversation at the conference: *Towards a Novel Assessment Framework for Cities and Climate Change*; *Urban Data Science for Global Climate Solutions*; *Responding to Climate Change in Cities and in their Informal Settlements and Economies*; and *Financing Low-Carbon, Climate-Resilient Cities*; and, *Urban Climate Change Science, Impacts and Vulnerabilities: State-of-the-Art Findings and Key Research Gaps* (Creutzig et al., 2019; UN-Habitat, 2018). The papers were commissioned primarily through professional connections of committee members, some of whom contributed as authors to these papers. Four of the five commissioned papers were included in the conference proceedings, with the fifth being developed while the proceedings were published (UN-Habitat, 2018, pp. 280–375).

A final piece of preparatory work before the 2018 conference was the *Nature* commentary, developed and published by 11 members of the Scientific Steering Committee (Bai et al., 2018). This article was developed as a provocation toward prioritizing low-carbon construction, alternative transport, and climate-resilient planning and design, and symbolized a starting point for the development of the research agenda; it was published in the week before the conference.

3.2.2 Methodology

The conference was convened by the two central committees, with the Scientific Steering Committee co-chaired by representatives of the IPCC, the scientific community, and the urban-focused community, with no overlap on the Organising Committee. "Information for the Research and Action Agenda was compiled from all sources of official conference inputs, which include plenary and parallel sessions, posters, and the five papers which were commissioned by the Scientific Steering Committee

(SCC) in advance of the Conference. A team comprised of representatives from partner organizations, two members of the SSC, and led by Future Earth was charged with compiling all conference inputs and supporting the SSC in the writing of the Global Research and Action Agenda on Cities and Climate Change Science" (IPCC, 2018, p. 54). At two points during the conference, members of the Scientific Steering Committee and the Organising Committee, along with other prominent attendees, such as plenary session speakers from the day, were invited to come together to discuss what had been heard over the previous 24 hours at the conference. The compilation team also met each evening of the conference to bring together the information collected from each of the day's sessions. This formed the basis for the initial discussion on the first day of the two-day post-conference workshop. Scientific Steering Committee and Organising Committee members attended the first day of the workshop; the second day was reserved for discussion among Scientific Steering Committee members.

The workshop concluded with a rough outline of the GRAA (Hunter et al., 2022). One of the primary resources on which the GRAA authors relied as they developed the details of the GRAA were questionnaires given to session leads and poster presenters, to be filled in after the presentation and discussions at the conference. These had specific questions focused on identifying the gaps relating to the areas of research presented in each session and on each poster (UN-Habitat, 2018). The information collected was analyzed and placed in a table, which was used throughout the GRAA development process to ensure that no central topics of discussion were absent from the text. Additionally, as the author team worked toward finalizing the text, the table was used as a reference to ensure that any research gaps that had been identified were explicitly discussed during the three days of the conference. This helped to reduce author bias by minimizing undue focus on specific topics.

The Scientific Steering Committee members, along with one Organising Committee member who had joined the GRAA author team, met in June 2018 in Mexico City to finalize the initial draft of the GRAA. During this time, the initial wheel diagram was also conceptualized as a visual representation to bring together all three sections of the 2018 GRAA. The group also developed a questionnaire to be sent to all conference participants along with the initial draft of the GRAA as a way for them to provide feedback on the initial draft, look for gaps, and ensure that no topics strayed beyond what was heard at the conference. The GRAA authors incorporated the feedback received on the initial draft into the final GRAA, which was submitted to the IPCC panel during its 48th session in October 2018 as part of the conference report.

3.2.3 Contributions and Perspectives

Over 700 participants from across academia, city leadership, the private sector, and civil society attended the inaugural conference, providing insights that informed and shaped the co-produced GRAA. Organizing partners included Cities Alliance, City of Edmonton, C40 Cities Leadership Group, Future Earth, ICLEI—Local Governments for Sustainability, IPCC, Sustainable Development Solutions Network, United Cities and Local Governments, United Nations Environment Programme, United Nations Human Settlements Programme (UN-Habitat), and the World Climate Research Programme. At the time, it was also noted that "the Global South—especially those working in Global South institutes—could have been better represented. Respondents appreciated the open dialogue between more-established professionals and academics and early-career participants" (IPCC, 2018, p. 5).

3.2.4 Findings and Outcomes

The GRAA was the primary outcome of the 2018 conference. It included three concentric rings to represent the various dimensions of consideration at the intersection of cities and climate science. The orange inner circle presented key cross-cutting issues and knowledge gaps for a step-change of knowledge generation on cities and climate change. The multi-colored middle circle presented six research areas where more evidence was deemed necessary to inform action. The green external circle presented three suggested approaches that could facilitate the agenda's implementation. The cross-cutting issues and topical research areas were more clear-cut areas of focus in 2018, when the GRAA was being developed. However, there was much discussion within the author team about the content of the third, outermost circle; particularly, whom it should represent and how it should direct and encourage engagement for action. The GRAA authors thought it very important for all urban stakeholders to see themselves in this level of support for action, and an early iteration on the GRAA had this section grouped by different levels of government that would be key in supporting action toward the outlined research and innovation priorities (GCoM, 2018a). However, the final version is broader and more inclusive. Conference participants saw the initial wheel diagram when they were sent the GRAA draft for review, but they did not influence the initial formulation of the diagram.

"The 358 individual research gaps and approaches for action which were identified and are categorized under the section headings found in the agenda" (IPCC, 2018, p. 56) were presented with section headings used as groupings for subsequent review, as "the breakdown of the key messages found in this table was developed through consultation that took place throughout the writing process" (IPCC, 2018, p. 56). While there was no decision at the conference, and the association between Innovate4Cities and the IPCC came later, carrying forward the intent of the Edmonton Declaration (GCoM, 2018b), on the day after the 2018 conference, in a conversation between the Scientific Steering Committee and the Global Covenant of Mayors for Climate and Energy (GCoM), it was suggested that support for the Innovate4Cities initiative would come from the GCoM, "to work with cities to further develop this Research and Action Agenda to align specifically with their needs" (IPCC, 2018, p. 15; Fig. 3.2).

3.3 Diving Deeper into City Dimensions: The 2021 GRAA and CRIA

3.3.1 Background and Context

Following the 2018 conference, the GCoM launched the Innovate4Cities initiative to support and expand the research and action called for within the 2018 conference's GRAA. The initial product emerging from this initiative was the City Research and Innovation Agenda (CRIA). This was developed through six workshops, which brought mayors and the private sector together with the academics, urban practitioners, and policymakers who had helped to shape the 2018 GRAA at the Cities and Climate Change Science Conference. GCoM membership grew from 9098 cities at the time of the 2018 conference to 11,714 by the 2021 Innovate4Cities Conference, demonstrating both the increased engagement and awareness across the multilateral local government landscape and the mounting diversity and complexity of delivering research and innovation action as delineated in the GRAA (GCoM, 2021, p. 9).

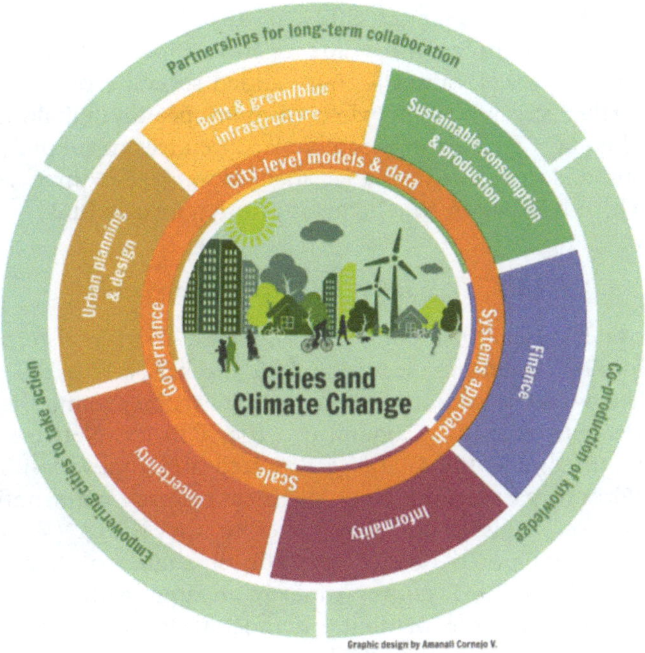

Fig. 3.2 The 2018 GRAA (*Source* Prieur-Richard et al., 2019, p. 3, used with permission)

3.3.2 Methodology

The fully online nature of the Innovate4Cities 2021 conference, guided by the Science and Innovation Steering Committee, expanded on the 2018 approach, which had established the model for participants making session contributions and established weighting systems for selection for the conference agenda (IPCC, 2024, pp. 13–14). This led to a rolling, global program schedule. Session chairs and presenters identified research gaps and areas of innovation associated with their session. Building upon the student-led report efforts of the 2021 Conference (University of Melbourne, 2021) student writers acted as rapporteurs and note-takers for all sessions, supplementing the session-leader questionnaire responses with a summary synthesized into the 2021 updated

GRAA (GCoM & UN-Habitat, 2022, p. 43). The Science and Innovation Steering Committee, with guidance from the GCoM Research and Innovation Technical Working Group, added several cross-cutting areas needing further research to the wheel diagram: health, digitalization, and justice and equity. They invited conference attendees to submit proposals for sessions and presentations that focused on these three additional themes, which had begun to emerge as priority areas in 2021.

3.3.3 *Contributions and Perspectives*

The Innovate4Cities 2021 conference involved 6901 registered participants from 159 countries, featuring 886 speakers across 191 sessions. During five working sessions involving the writing team, the Science and Innovation Steering Committee, the Partner Network Organising Committee, and the GCoM's Research and Innovation Technical Working Group, participants recorded their research contributions and perspectives for inclusion in the updates to the GRAA (Walsh et al., 2022, p. 43) and noted down any shifts or additional areas of focus in Innovate4Cities 2021 conversations compared with those held in 2018. The potential role of digital democratization and decentralized participation explored in the 2021 virtual conference demonstrated a climate-responsive means of convening a more diverse body of participants, but the potential for transformative change was thwarted by draconian COVID quarantine restrictions.

3.3.4 *Findings and Outcomes*

The key takeaways from the Innovate4Cities 2021 Conference included the prioritization of equity and justice; taking a systems approach; examining mental models and taking an end-to-end approach to urban climate science that would consider the speed and scale of activities, including knowledge-gathering, socio-cultural dimensions of climate change, and communication and collaboration in creating a safe space; finding allies and role models; and closing the implementation gap (GCoM & UN-Habitat, 2022, p. 11). This was reflected specifically in the 2021 GRAA by the addition of a framing section before identifying research gaps in cross-cutting and topical areas. A triple-helix illustration was also added to illustrate the fact that the conference discussions could not be fully encompassed in the wheel diagram. This additional diagram was meant to illustrate the key aspects that would be needed to catalyze implementation to ensure that climate action in cities occurred at the required speed and scale. Additionally, a further topical research area emerged from conference discussions: focused on history and cultural heritage. Separately, the City Research and Innovation Agenda succinctly summarized the expansion of the 2018 GRAA, encapsulating the proceedings of the Innovate4Cities 2021 Conference (UN-Habitat, 2022).

The 2021 GRAA and CRIA were released and presented at the 2021 ICLEI World Congress in Malmo, Sweden. A workshop session was held specifically to encourage stakeholders to further explore the topics of history and cultural heritage and to encourage further collaboration and research. Inspired by this session, Ives and Baker (2024) developed an intelligence brief that presented some of the arguments as to why faith perspectives are vital to enabling (or constraining) climate action and proposed a model of multi-actor partnership across religious and secular divides based on shared values. In the 2021 GRAA/CRIA document, an understanding was reached that "to close the implementation gap, new research and knowledge on methods to change the mindset of urban stakeholders around addressing climate change and the tools and processes to do so...came through very strongly as needs throughout the Innovate4Cities 2021 conference" (GCoM & UN-Habitat, 2022, p. 6). Participants recognized that the wheel structure was inadequate to represent the research gaps; this was initially addressed with the triple-helix design to move beyond the reductive elements of two dimensions (Fig. 3.3).

Diagramming an ever-evolving understanding of the increasing complexity of urban action and research—a challenge that became clear during the 2021 GRAA/CRIA formulation process-became the core of preparing for the 2024 GRAA.

3.4 A Unified Structure: The 2024 GRAA

3.4.1 Background and Context

Following the 2021 GRAA updates, the 2024 restructuring sought to answer the following questions: "What does mainstreaming the GRAA in the research and innovation lexicon of climate and cities-focused research look like from 2024 onward?" and "How may Innovate4Cities strengthen the multidisciplinary approach to urban research and innovation with a

Fig. 3.3 The 2021 GRAA (*Source* GCOM & UN Habitat, 2022, p. 20)

cross-dimensionally aligned, updated GRAA structure?" (GCoM, 2024a, p. 42). Key to answering these questions was the effort to align the broader thematic dimensions of the GRAA with the increasingly detailed and interconnected dimensions of the CRIA. While both structured agendas had been refined in 2021, the linkages between the two had not necessarily been drawn together in a manner that allowed representation of each agenda in relation to the other.

3.4.2 Methodology

The process of synthesizing the 2021 GRAA and CRIA into a single framework involved a textual analysis of the conceptual overlap between the two documents, identifying which shared language appeared with the greatest frequency to determine the emphasis required in the updated 2024 edition. A review of abstracts submitted to the conference—296 reviewed, with 254 session data sheets collected—were combined with abstracts and feedback from the proceedings and juxtaposed with key multilateral framework documents against the new GRAA for a quantitative textual analysis of how GRAA dimensions and affiliated language appeared in each text (GCoM, 2024a, 2024b, p. 44). These findings helped determine the research gaps and priority actions to populate the various dimensions of the GRAA.

3.4.3 Contributions and Perspectives

With nearly 2,000 in-person and online participants, the 2024 Innovate4Cities conference represented the most sectorally diverse and balanced body of participants up to that time. A marked increase in civil-society participation was noted, as the relevance of contributions to climate-resilient development from beyond the conventional peer-reviewed academic avenues and public-sector policy preparation was being recognized. Bridging between sectors, making connections, and building a shared vocabulary beyond specialist terminology had become a critical component of addressing dimensions of justice and equity otherwise underserved by a narrow-band dialogue partitioned along disciplinary lines.

The linkage of graphic representation and semantic value has been a valuable tool for facilitating translation of concepts and language (Piccinini, 2010, p. 99); thus, the restructuring of the GRAA was undertaken with the benefits of spatial mnemonics in mind. While the two-dimensional wheel diagram offered a presentation of the various topics involved in city-scale climate science, the structural form introduced in the 2024 iteration of the GRAA established the potential to organize research and innovation contributions in terms of their relationships to the relevant dimensions. Input throughout the 2024 conference led to a refinement and expansion of the dimensions to include additional topical and cross-cutting areas previously either subsumed under or left unaddressed or unacknowledged by the broader categories previously presented (for example, decentralization and uncertainty as dimensions of city-level models, data, and knowledge).

3.4.4 *Findings and Outcomes*

The GRAA can now explain the theoretical and methodological approaches being applied across the various spaces of local-level research, regardless of disciplinary positioning (GCoM, 2024a, 2024b, p. 49). The *160 research gaps* and *259 priorities for action* identified as of the 2024 GRAA update build upon those included in the 2018–2021 GRAA (GCoM, 2022) and 2021 CRIA (p. 50). The 2024 redesign of the GRAA now provides a visual map of research, orienting and connecting its constituent dimensions. In its new form, with the contributions and perspectives of participants taken into account, the GRAA is better positioned to inform the IPCC SRCCC through clearer relationships across the various areas of the agenda (Figs. 3.4, 3.5).

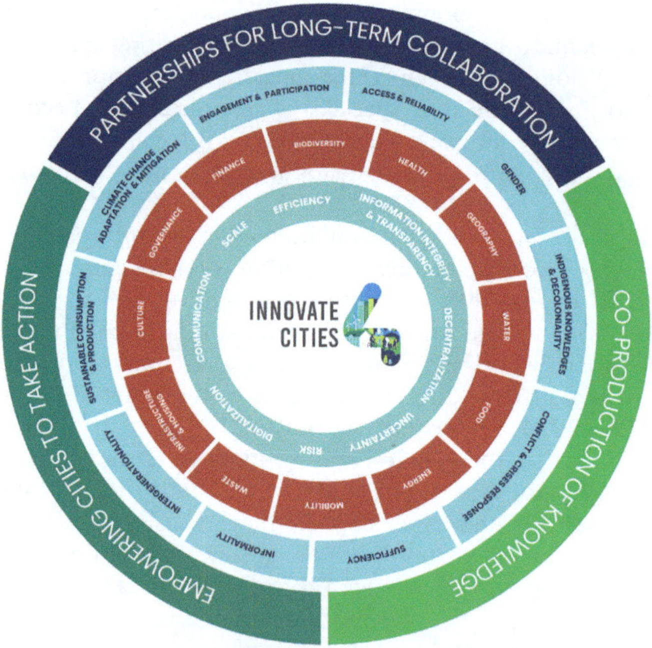

Fig. 3.4 The 2024 GRAA (*Source* Irvin et al., 2024, p. 45, Image created by Genov, N.)

Fig. 3.5 The 2024 GRAA structured reconfiguration (*Source* Irvin et al., 2024, p. 48, Image created by Genov, N.)

3.5 Looking Ahead: Reflections on the Past to Inform the Future—Connecting Research to Action Through the GRAA and Facilitating Dynamic Approaches to Urban Climate Governance

3.5.1 *Evolving Role of GRAA*

Future iterations of the GRAA will strengthen connections across the levels, with partners for implementation (i.e., cities, other levels of government, industry, and research institutes) encouraged to identify their roles in addressing the research gaps and priorities for action. Eliciting this participation via sector-specific briefings will help further populate the digital rendition of the GRAA, which is to serve as a navigable data directory that catalogues contributions to the Innovate4Cities initiative for accessibility to urban-focused researchers and practitioners looking for information on the dimensions of relevance to their immediate urban challenges. The GRAA is intended to facilitate municipalities' efforts to tie their city journey to dimensions of climate-resilient development, and to prompt questions about causes and effects of policies and activities. The use of the GRAA for mapping municipal policies, plans, and actions provides a structure for visualizing their research and innovation activities, making it easier to communicate case studies to multi-level governance partners (and other sectoral stakeholders).

In the global response to climate change, multilateral and transnational organizations have traditionally relied on state-led, compliance-oriented strategies. These strategies—referred to here as *static climate governance*—focus on emissions targets, standardized reporting, and technical solutions aimed at controlling current sources of emissions (UNEP, 2017; Green, 2014; Hickmann, 2015). While previous approaches of sovereign state-scale multilateralism have laid foundational groundwork for international cooperation (Depledge & Yamin, 2009; Karns & Mingst, 2009), they often overlook the social complexity, uncertainty, and adaptive challenges that define urban climate action in complex, multi-actor scenarios by failing to engage with the community-level realities of national-level policy instruments.

Cities today face increasingly multifaceted and urgent climate threats, underscoring the need for governance models that are flexible, inclusive, and capable of stimulating bottom-up innovation. In response, the

GRAA has evolved to reflect a paradigm shift toward dynamic, human-centered approaches (Hale, 2016; Mission Innovation, 2021). As the foundation of the GRAA is the empowerment of cities to take action, the pillars of justice and equity now identified as supporting dimensions of multi-system functionality represent the greatest expansion of scope and consideration in the 2024 GRAA. City-level models, data, and knowledge are now properly recognized as dimensions of system interaction and are positioned as information tools for delivering justice and equity for all residents of urban environments.

Traditional transnational initiatives (for example, the UN's Clean Development Mechanism and the International Civil Aviation Organization's CORSIA market-based measures) have typically pursued direct or "static" climate solutions that are focused on regulating existing emissions sources through mechanisms such as carbon offsetting, standard-setting, and monitoring (ARUP, 2015; Green, 2014; ICAO, 2025). The range of climate response interventions proposed through resource mobilization mechanisms over recent decades have been critical to operationalizing international commitments, such as the Kyoto Protocol and the Paris Agreement (United Nations, 2015). However, one critique of established approaches is that they rest on the assumption that systems and actors will remain stable over time (Mission Innovation, 2021). They often fail to fully use cities' capacities to generate creative solutions aligned with changing human needs and consumption behaviors (Assarkhaniki et al., 2022b), such as the rapidly changing resource-access requirements of digital communication or compensatory actions required to respond to interrupted supply chains due to hazards such as pandemics, floods, and droughts.

Dynamic approaches, by contrast, reframe urban stakeholders not simply as emitters of greenhouse gases and other pollutants, but as generators of solutions. This involves focusing on indirect levers of change—such as infrastructure use, food systems, energy behavior, and cultural values—to address emissions through shifts in societal norms and everyday practices (Hale, 2016; Hickmann et al., 2021). One relevant example is transitioning food-system responses from decarbonizing meat production to reducing food waste, improving nutritional awareness, and rethinking dietary preferences (Mission Innovation, 2021). The restructured GRAA clarifies the ways interventions affect different dimensional spaces of consideration across systems.

3.5.2 *How the Innovate4Cities Network Intends to Use the GRAA*

Going beyond the periodic update undertaken upon the convening of Innovate4Cities conferences, the implementation of the GRAA still requires the formation of a virtual agglomeration of practitioners across the city/climate science community. In a digitized library where resources may be uploaded, indexed, and circulated freely across the network and beyond, resources mapped within the GRAA structure may inform dedicated practitioners' responsive solutions to multidimensional problems the world over. The development of an indexed, searchable database of GRAA-related research is expected to provide a means for Innovate4Cities participants across the network to engage with the research more directly as the body of published work at the intersection of urban climate science continues to grow; ideally, this would include digital fora to discuss material between formal conference events.

3.5.3 *Additional Contributions and Perspectives Needed*

A key necessity, as practitioners in municipalities globally engage with the dimensions of the GRAA, is the translation of concepts of "a systems approach to research and innovation in urban planning and design utilizing city level models to deliver just and equitable action in climate resilient development" (GCoM, 2024a, p. 47). Mapping these concepts across cultures and languages is a necessary step to connect the dimensions of research and innovation generally recognized by the Innovate4Cities community of practice at the local level to terminology used in communities around the world. Syntax and structure in the English presentation of the GRAA only go as far as they are understood and communicated; to receive iterative feedback inclusive of other perspectives, practitioners need tools to relate these ideas across the sectors of their own localities in the simplest, most effectively understood means available. A mechanism for continual peer review through an expanding network of researchers and climate-resilient development practitioners across the world can encourage cross-cultural comparison and improvements in both knowledge fidelity and its transmission.

3.5.4 Gaps and Actions: Targets and Goals

Despite cultivating an understanding of the city-climate interface that captures the scope of global challenges with increasing complexity, the research gaps and priority actions identified and revised in each iteration of the GRAA serve as targets and goals for attainment, both at the individual municipal scale and at the level of global responses.

In this context, the GRAA is well-positioned to serve as a bridge between traditional multilateral frameworks and emerging innovation ecosystems. The development of a "City Research and Innovation Bridge" framework under the GRAA seeks to align city-identified knowledge and capacity gaps with dynamic, need-based climate solutions to link priorities across levels of governance and sectors (Assarkhaniki et al., 2022a; Oke et al., 2022).

This evolving role is particularly relevant given the scope of participation already mobilized through global networks. More than 12,000 cities are now part of the GCoM, with around 6000 adopting local climate action plans (GCoM, 2021). Yet many cities still face barriers to implementation, such as limited access to localized data, tailored technologies, and adaptive policy mechanisms (GCoM & UN-Habitat, 2022). The GRAA framework, through its iterative refinement and collaborative design, helps reveal these barriers and address them in the context of global innovation efforts.

The path forward will require future iterations of the GRAA to further embrace these dynamic dimensions. Doing so means integrating flexible innovation logic throughout its structure (i.e., recognizing how changes in dimensions of justice and equity or city-level models, data, and knowledge across systems may be linked, replicated, and/or adapted to other systems), enabling solution-providers, researchers, and local governments to co-create responses that reflect both scientific rigor and social relevance. As a unifying platform, the GRAA offers not just a catalog of gaps and priorities, but a strategically structured interface for turning city-level ambition into global implementation through highlighting areas of need and opportunities to act across interrelated dimensions of society's various systems.

3.6 Conclusion

3.6.1 GRAA as a Measurement Tool for State of the Science

The multidimensionality of the GRAA provides an opportunity to evaluate the contributions of the community of practice across disciplines toward addressing climate-related issues through city-scale interventions. As shown by the evolving participation in the Innovate4Cities conferences, the community of practice is steadily broadening beyond academia and local government. Those practitioners with findings pertaining to specific subjects (for example, intergenerationality and food systems) may be framed against other dimensions (such as scale, conflict and crisis response, and transport systems), expanding the scope of inquiry to better understand how the research can be built upon through further investigation, interrogating findings, and providing a basis for iterative improvements to ongoing and future research across all dimensions.

3.6.2 GRAA as a Responsive Reflection of Urban Needs

Because the intervening period between conferences leaves room for activities that go beyond supplementary regional events and Innovate4Cities marketplace activities to elicit greater multi-sectoral engagement around the world, a persistent platform for climate science discourse between practitioners in cities around the world may provide space for responsive reflection from the participating community about how initiatives may be adapted and replicated elsewhere (GCoM, 2024b). The GRAA provides a blueprint for structuring a forum for ongoing dialogue across relevant topics of research and action.

3.6.3 GRAA as a Rhetorical Tool for Instigating Action

The research gaps and priority actions provide a benchmark for motivated, coordinated effort, both at a local level and aggregated toward a global understanding of how cities are responding to climate change. The spatial mnemonics of the GRAA structure provide a means of visualizing where action is taking place, and where gaps remain to be addressed based on information from practitioners across the city-climate science landscape. The GRAA structure allows users working in areas like finance, farming, and wastewater treatment to trace intersections and linkages between activities across systems and their related dimensions.

3.6.4 GRAA as a Model for Motivating Decentralized Research and Action

Across All Levels

The initial intent of the GRAA to serve as a guidance framework that complements those previously developed at national and international levels (for example, the Sustainable Development Goals, UNFCCC Nationally Determined Contributions, and the Convention on Biological Diversity) is increasingly relevant, particularly as opportunities for improved efficiency and efficacy in climate response slip past those governing at their respective levels without cohesive coordination. The GRAA may provide a model for instigating multi-level partnerships from the bottom-up.

Across All Sectors

Laying a foundation for empowering cities toward delivering justice and equity in climate-resilient development establishes a precedent for inclusive decision-making and participatory research and action: not only by academics, but by all sectors of urban life, including marginalized and vulnerable community members otherwise poorly included in climate policy efforts.

Acknowledgements The authors would like to thank those participants that shaped the GRAA over its previous three iterations.

Competing Interests

The authors have no conflicts of interest to declare that are relevant to the content of this chapter.

References

ARUP. (2015). *Potential for climate action. Cities are just getting started.*

Assarkhaniki, Z., Oke, C., Walsh, B., Jance, B., Pamlin, D., & Deacon, A. (2022a). *City research and innovation agenda bridge—Dataset*. https://doi.org/10.26188/21399585.v4

Assarkhaniki, Z., Oke, C., Walsh, B., Jance, B., Pamlin, D., & Deacon, A. (2022b). *City research and innovation agenda bridge*. https://doi.org/10.26188/21405642.v2

Bai, X., Dawson, R., Ürge-Vorsatz, D., Delgado, G., Barau, A. S., Dhakal, S., Dodman, D., Leonardson, L., Masson-Delmotte, V., Roberts, D. C., & Schultz, S. (2018). Six research priorities for cities and climate change. *Nature*, *555*(7694). https://doi.org/10.1038/d41586-018-03409-9-z

Creutzig, F., Lohrey, S., Bai, X., Baklanov, A., Dawson, R., Dhakal, S., Lamb, W. F., McPhearson, T., Minx, J., Munoz, E., & Walsh, B. (2019). Upscaling urban data science for global climate solutions. *Global Sustainability, 2*, Article e2. https://doi.org/10.1017/sus.2018.16

Depledge, J., & Yamin, F. (2009). The global climate-change regime: A defence. In D. Helm & C. Hepburn (Eds.), *The economics and politics of climate change*. Oxford University Press.

GCoM & UN-Habitat. (2022). *Discussions and findings from the 2021 Innovate4Cities conference: New insights for the 2018 global research and action agenda on cities and climate change science*. https://doi.org/10.26188/62795d16f3bab

Global Covenant of Mayors for Climate & Energy. (2018a). *Innovate4Cities—A global climate action accelerator: Research and innovation priorities* (OPS Booklet). https://www.globalcovenantofmayors.org/wp-content/uploads/2018/09/GCoM_Innovate4Cities-OPS_Booklet_8.5x11.pdf

Global Covenant of Mayors for Climate & Energy. (2018b). *Innovate4Cities—A global climate action accelerator—Edmonton declaration*. Edmonton, Alberta, Canada for the Change for Climate—Global Mayors Summit. 3–4 March, 2018. https://www.globalcovenantofmayors.org/wp-content/uploads/2018/05/Edmonton_Declaration_Update_May23_v2.pdf

Global Covenant of Mayors for Climate & Energy. (2022). *City research and innovation agenda*. Innovate4Cities. https://www.globalcovenantofmayors.org/wp-content/uploads/2022/05/GCoM_Innovate4Cities_2022_Booklet.pdf

Global Covenant of Mayors for Climate & Energy. (2024b). *I4C marketplace resource library*. Innovate4Cities. https://www.globalcovenantofmayors.org/resource-library-search/#i4c-marketplace

Global Covenant of Mayors for Climate & Energy. (2024a). *Shared spaces for urban innovation: Insights and outcomes from Innovate4Cities marketplaces*. Innovate4Cities. https://www.globalcovenantofmayors.org/wp-content/uploads/2024/02/I4C-Marketplaces-Outcomes-Insights-Report_.pdf

Global Covenant of Mayors. (2021). *Further and faster together: The 2021 global covenant of mayors impact report*.

Green, J. F. (2014). Rethinking private authority: Agents and entrepreneurs in global environmental governance. *Princeton University Press*. https://doi.org/10.1515/9781400848669

Hale, T. (2016). "All hands on deck": The Paris agreement and nonstate climate action. *Global Environmental Politics, 16*(3). https://doi.org/10.1162/GLEP_a_00362

Hickmann, T. (2015). *Rethinking authority in global climate governance*. Routledge.

Hickmann, T., Widerberg, O., Lederer, M., & Pattberg, P. (2021). The United Nations framework convention on climate change secretariat as an orchestrator in global climate policymaking. *International Review of Administrative Sciences, 87*(1), 21–38. https://doi.org/10.1177/0020852319840425

Hunter, N., Roberts, D., Sutherland, C., & Slotow, R. (2022). Co-creation between cities and climate change science achieves research and action agenda. *Current Research in Environmental Sustainability, 4*, Article 100189. https://doi.org/10.1016/j.crsust.2022.100189

ICAO. (2025). Status of CORSIA implementation. *International civil aviation organization*. https://www.icao.int/sites/default/files/environmental-protection/CORSIA/Timeline-CORSIA-framework-maintenance.pdf

IPCC. (2018). *Progress report—International conference on climate change and cities—Forty-eighth session of the IPCC Incheon, Republic of Korea, 1–5 October 2018*. IPCC-XLVIII/INF. 1, Rev. 1 (10.IX.2018) Agenda Item: 4.1. https://www.ipcc.ch/apps/eventmanager/documents/52/100920181041-INF1Rev1CitiesReport.pdf

IPCC. (2024). *Report of the Innovate4Cities conference in 2021—Sixty-first session of the IPCC Sofia, Bulgaria, 27 July-2 August, 2024*. IPCC-LXI/INF. 12 (9.VII.2024) Agenda Item: 13.1. https://apps.ipcc.ch/eventmanager/documents/87/120720241044-INF.%2012%20-%20AOB_I4Cities_IPCC_report_final.pdf

Irvin, A., Oke, C., Khan, S., Jance IV, B., Ranalder, L., Barth, B., Deacon, A., Abusamra, R., Addison, A., Adebisi, J., Adsett, C., Ailio, E., Al Jadaa, D., Aslan, S., Assalini, S., Assarkhaniki, Z., Astolfo, J. P., Badino, M., Bai, X., ..., & Williams, A. (2024). Next-generation city climate action: Findings from the 2024 Innovate4Cities conference and update to the global research and action agenda on cities and climate change science (GRAA). Global Covenant of Mayors for Climate and Energy, UN-Habitat & University of Melbourne. https://doi.org/10.26188/27147612

Ives, C. D., & Baker, C. (2024). Engaging faith for a sustainable urban future. *Global Sustainability, 7*, Article e45. https://doi.org/10.1017/sus.2024.32

Karns, M. P., & Mingst, K. A. (2009). *International organizations: The politics and processes of global governance*. Lynne Rienner Publishers.

Mission Innovation. (2021). *21st Century climate innovation assessment*.

Montfort, S., Callaghan, M., Creutzig, F., Lamb, W. F., Lu, C., Repke, T., Ge, K., & Minx, J. (2025). Systematic global stocktake of over 50,000 urban climate change studies. *Nature Cities, 2*, 613–625. https://doi.org/10.1038/s44284-025-00260-8

Oke, C., Walsh, B., Assarkhaniki, Z., Jance, B., Deacon, A., & Lundberg, K. (2022). The city research and innovation agenda: Prioritizing knowledge gaps and policy processes to accelerate city climate action. *Journal of City Climate Policy and Economy, 1*(1), 94–110. https://doi.org/10.3138/jccpe-2022.1.1.0007

Piccinini, C. (2010). The Xiguo jifa. Treatise on Western mnemonic arts by Matteo Ricci S.J. Its importance in Chinese language learning. In R. Malek & G. Criveller (Eds.), *Light a Candle. Encounters and friendship with China. Festschrift in Honour of Angelo S. Lazzarotto P.I.M.E.* (pp. 99–114). COLLECTANEA SERICA.

Prieur-Richard, A.-H., Walsh, B., Craig, M., Melamed, M. L., Colbert, M., Pathak, M., Connors, S., Bai, X., Barau, A., Bulkeley, H., Cleugh, H., Cohen, M., Colenbrander, S., Dodman, D., Dhakal, S., Dawson, R., Espey, J., … & Ürge-Vorsatz, D. (2019). *Global research and action agenda on cities and climate change science*. Intergovernmental Panel on Climate Change. https://www.ipcc.ch/site/assets/uploads/2019/07/Research-Agenda-Aug-10_Final_Long-version.pdf.

UNEP. (2017). *The emissions gap report 2017*. United Nations Environment Programme. https://www.unep.org/resources/emissions-gap-report-2017

UN-Habitat. (2018). *Cities and climate change science conference proceedings*. UN-Habitat. https://unhabitat.org/sites/default/files/2021/03/Cities%20IPCC%20Proceedings%20FINAL%20for%20Email-S.pdf

UN-Habitat. (2022). *Innovate4Cities 2021 conference proceedings*. Innovate4Cities. https://unhabitat.org/sites/default/files/2022/09/i4c_conference_proceedings.pdf

United Nations. (2015). *Paris agreement*. https://unfccc.int/sites/default/files/english_paris_agreement.pdf

University of Melbourne. (2021). *Innovate4Cities conference 2021: student-led final report*. Innovate4Cities. https://figshare.unimelb.edu.au/ndownloader/files/31907201

Walsh, B., Greenwalt, J., Oke, C., Hadfield, P., Dickey, A., H Craig, M., North, M., Hunter, N., Okem, A., Bland, S., Verbeeck, J., Barth, B., Deacon, A., Roberts, H., Randhawa, J., Joshi, S., Silva, D. C., Lootens, M., Adler, M., …, & Zhu, S. (2022). *Findings from Innovate4Cities 2021 and update to the global research and action agenda (Version 3)*. The University of Melbourne. https://doi.org/10.26188/19624029.v3

CHAPTER 4

Boundary Spanning for Climate-Responsive Cities

Sareh Moosavi, Bridget Keane, Jorn Verbeeck, Judy Bush, Debra C. Roberts, and Cathy Oke

4.1 Introduction: Boundary Spanning for Climate Action

Responding effectively to the multifaceted challenges of climate change, biodiversity loss, and the Sustainable Development Goals (SDGs) in urban contexts requires new, integrative modes of practice in decision-making. Traditional city-governance approaches—dominated by discipline-specific, sectoral, and isolated strategies—have consistently

S. Moosavi (✉) · B. Keane · J. Bush
Faculty of Architecture Building and Planning, University of Melbourne, Melbourne, Australia
e-mail: sareh.moosavi@unimelb.edu.au

B. Keane
e-mail: bridget.keane@unimelb.edu.au

J. Bush
e-mail: judy.bush@unimelb.edu.au

C. Oke et al. (eds.), *A Global Research and Action Agenda on Cities and Climate Change Science*,
https://doi.org/10.1007/978-981-95-6517-7_4

fallen short addressing these complexities. These approaches often lead to fragmented or mono-functional solutions that lack coordination across scales, sectors, and stakeholders, failing to address the spatial and social complexities inherent in urban systems (Eriksen et al., 2024).

Given the dynamic and interconnected nature of urban socio-ecological systems, ongoing collaboration between science, policy, and practice is essential to ensure an integrated approach. This requires co-producing knowledge across these domains and fostering processes that enable this integration. One conceptual framework with foundations in social sciences that has gained traction in explaining knowledge transfer processes is boundary work (Edwards, 2011). Boundary work theory offers valuable insights into how knowledge is transferred and translated at boundaries between scientific research, policymaking, and on-the-ground implementation (Edwards, 2011). Clark et al. (2010) define boundary work as the set of strategies, processes, and institutions that help mediate between different communities of knowledge and practice. They argue effective boundary work should maintain the integrity of each domain, where science remains credible, policy remains legitimate, and decisions are seen as salient, to create a shared space where knowledge can be co-produced and translated into action.

In the context of urban climate action, the concept of "boundary spanning" offers an anchor to investigate the activities or tactics that

---

J. Verbeeck
Dark Matter Labs, Antwerp, Belgium
e-mail: jorn.verbeeck@gmail.com

D. C. Roberts
School of Life Sciences, University of KwaZulu-Natal, Durban, South Africa

Faculty of Geo-Information Science and Earth Observation, University of Twente, Enschede, The Netherlands

D. C. Roberts
e-mail: Debra.Roberts@palebludot.co.za

C. Oke
Melbourne Centre for Cities, Faculty of Architecture Building and Planning, University of Melbourne, Melbourne, Australia
e-mail: cathy.oke@unimelb.edu.au

are needed to bridge the agendas and priorities of actors separated by various types of boundaries (Dąbrowski, 2017). "Knowledge boundary spanners"—also referred to as boundary actors—have been recognized as critical intermediaries (Dąbrowski, 2017) who play a vital role in interpreting complex climate science and place-based expertise (including culturally nuanced forms of climate science) and converting them into actionable policies and interventions suited to the urban environment. The terms "boundary spanner" and "intermediary" are used in different ways by different research communities, and there is a lack of agreed definitions (Neal et al., 2023). Boundary spanners (Burbach et al., 2023) and intermediaries (Fischer & Newig, 2016) can be understood as working between and linking different policy domains and levels of government. Their work goes beyond simple knowledge communication, which distinguishes them from knowledge brokers. Knowledge brokering is usually applied to the process of disseminating scientific research into policy-making processes (Karcher et al., 2025). Neal et al. (2022) distinguish between intermediaries as organizations engaged in research dissemination, and boundary spanners as individuals involved in facilitating relationships. Boundary spanners actively reconcile divergent epistemologies, timeframes, and priorities that distinguish academic science from urban governance systems (Tribbia & Moser, 2008), while navigating power and practicing ethics in translating knowledge (Gibbs & Ross, 2025).

Urban climate actions that work towards climate resiliency in cities demand integrated mitigation and adaptation strategies that are harmonized across different levels of governance, sectors, actors, and life stages of the built environment (Hürlimann et al., 2021, 2022). To achieve this integration while recognizing the plurality of knowledge systems, it is essential to foster dialogues that connect scientific research with other knowledge systems, including local, practical, tacit, and Indigenous knowledge systems (Hatch et al., 2023; Taylor et al., 2021). This calls for bridging actors and organizations who can advance knowledge boundaries.

While the significance of boundary spanning has been widely acknowledged among scholars in fields of administration, management, social innovation, and leadership (Cross et al., 2013; Hawkins et al., 2012; Satheesh et al., 2023), it has been less explored in built environment scholarship. In recent years, the relevance of boundary work has been highlighted in environmental management, sustainability, and infrastructure governance literature (Posner & Cvitanovic, 2019), where boundary spanning is increasingly recognized for shaping complex science–policy landscapes, as a means of cultivating trust, aligning priorities among diverse stakeholders, and building inter-organizational linkages (Dąbrowski, 2017; Goodrich et al., 2020). Boundary spanning activities were studied in large landscape-infrastructure projects by van Dijk et al. (2023) and van den Brink et al. (2019), highlighting different profiles of boundary actors and types of boundaries. Similarly, Moosavi et al. (2025a, 2025b) emphasized the role of landscape architects in advancing knowledge boundaries for transformative climate actions through integrating nature in cities and facilitating knowledge co-production between multiple stakeholders.

Despite this growing recognition, the specific dynamics of boundary spanning within the built environment—especially in relation to climate action in cities—remain understudied. Addressing this gap is particularly critical given that urban infrastructure, including buildings, transport networks, and even public open spaces (Moosavi et al., 2025a, 2025b), can be significant contributors to greenhouse gas emissions, while also being exposed to climate impacts (IPCC, 2022). Importantly, insights from practice on how boundary spanners operate across different roles and scales are not widely reported in the literature.

This chapter explores how different boundary spanners across private and public sectors operate in their bridging roles to address the science–policy–practice gaps for climate actions by focusing on three case studies of organizations and individual boundary spanners; a small private consultancy, a large intermediary organization, and a regional local government. We interviewed three key actors with questions that revolved around four key themes to identify; (a) what climate *actions* they mobilize, (b) what *types of boundaries* they span, (c) across which *life stages of built environments*, and (d) under what *enabling/disabling conditions*. Before outlining the methods (Sect. 4.3) and discussing the results (Sect. 4.4), we establish the identified types of boundaries discussed in the literature in the following section.

4.2 Types of Boundaries in Science–Policy–Practice Exchanges

Descriptions of boundary types across the literature vary depending on disciplinary areas of research and subject matter. In studies focusing on organizational management, boundary spanners are often seen to employ activities to cross organizational boundaries, including institutional and sectoral divides (Cross et al., 2013; Dąbrowski, 2017; Satheesh et al., 2023). Satheesh et al. (2023) divide these types of boundaries into vertical (e.g., boundaries between levels of government) and horizontal (e.g., boundaries between policy sectors), with Delozier and Burbach (2021) adding the temporal dimension, which concerns the boundaries between different time horizons (e.g., short- vs. long-term perspectives). Cross et al. (2013) identify three additional boundary types: stakeholder (partners, constituencies, value chain, communities, rightsholders), demographic (gender, age, nationality, culture, ideology), and geographic (location, region, markets).

At the intersections of environmental science and spatial management, boundary types acquire an expanded definition to include physical and material boundaries (e.g., structures that mark territories) (van den Brink et al., 2022), boundaries between project life stages (Moosavi et al., 2025a, 2025b), and knowledge domains (syntactic, semantic, pragmatic) (Carlile, 2004). Spanning each type of boundary poses different challenges. Table 4.1 summarizes the identified categories of boundary types and associated challenges, based on a synthesis of the literature. These categories were used to identify case studies that operate at different scales and across different boundary types, with a focus on the boundaries across project life stages. Boundary types further informed the interview questions and the analysis of the study results.

4.3 Methods: A Framework for Assessing Boundary Spanning in Practice

We identified three case studies of practices across private and public sectors that are operating at the intersections of science–policy–practice, leveraging the skills and capabilities of individuals who act as connectors, bridging knowledge domains to push climate agendas in the built environment. The case studies look at three organizations (and individuals) in the private and public sectors in three different countries; Dark Matter

Table 4.1 Boundary types and challenges

Boundary type	*Examples*	*Challenges*
Vertical	Institutional, governance (levels), power dynamics	Power imbalances (Delozier & Burbach, 2021)
Horizontal	Organizational, sectoral (e.g., policy sectors), disciplinary, knowledge systems, cultural	Language, alignment, trust (Cross et al., 2013; Posner & Cvitanovic, 2019)
Spatial	Geographical, physical, material, administrative, scalar	Scale mismatches, geo-politics (van Dijk et al., 2023)
Temporal	Time horizons, age, short-term vs. long-term acting times, project life stages	Evolving belief systems and values, competing timeframes (Delozier & Burbach, 2021; van den Brink et al., 2022)
Socio-cultural	Demographical, ideological, values, gender, communities	Conflicting values, worldviews, and priorities (Gibbs & Ross, 2025)

Labs in Denmark, Oculus Studio in Australia, and Menjez Municipality, a local council in Lebanon. Individuals interviewed have been de-identified to protect anonymity and support the research data collection process. Through semi-structured interviews with selected individuals from each organization, we asked about the following key areas:

- Their role and activities as boundary spanners;
- The types of boundaries they cross for climate actions;
- Where they often operate across life stages of projects; and
- What challenges they encounter or what opportunities they identify.

The boundary types presented in Table 4.1 provided a guide to categorize the types of boundaries that these individuals span in facilitating climate actions, which emerged from the interviews. During the interviews, participants were asked about their actions across the eight built environment life stages developed by Hürlimann et al. (2021): change initiation, strategic planning, project consideration, design, costing and approvals, construction, use and management, and recovery and renewal, along with 16 priority climate actions across these stages (Hurlimann et al., 2025). These actions were used to prompt the interviewees to also identify additional actions specific to boundary spanning at different stages. The findings are discussed in the following section.

4.4 Results: Boundary Spanning in Practice

4.4.1 Case Study 1: Private-Sector Design Consultancy, OCULUS Studio, Australia

Boundary Roles

The interview for this case study was conducted with the leader of a mid-scale landscape architecture and urban-design practice, OCULUS. With approximately 45 staff in three locations in Australia and one in New Zealand, the practice designs and delivers projects across most scales of landscape architecture (parks, gardens, education and health landscapes, public and institutional landscapes). In response to a question about their role in the context of boundary spanning and the boundaries that they cross, the interviewee first noted that they would not describe themselves as a boundary spanner, as they saw their role as "fitting firmly" within the professional designation of landscape architect. However, with further elaboration, it emerged that the ways in which that role was performed operated across boundaries through knowledge transfer and policy formation. Describing their activities, the practice leader discussed a series of actions explained as "stitching" and "looping." Stitching was a way of working with multiple disciplines as well as relating the design brief to other agendas.

> Each time you do those pieces of work, there's a piece missing in the brief. Each of them is asking the same question in a different way. I think for me it's stitching, of trying to stitch it all together.

Stitching was also seen as a way of contributing back to the client or commissioning body by drawing in research, tools, and ideas from a broader context.

> They've done other pieces of work, they've commissioned other research, they're very focused on their particular strategic needs or their budgetary constraints. We try and bring what we've already learned from other contexts into that.

Looping back was described as a process that followed stitching, to improve accountability and action within the practice for future projects, with the aim of complementing and providing evidence for actions that enhanced the work of the practice related to climate action.

Types of Boundaries

The interviewee noted types of boundaries crossed for transformative climate actions as predominately vertical, related to governance. This emerged in the context of discussing a process of moving beyond the scope of design services to engage with governance, which involved proactively contributing to policy debates and working with instruments for policy influence and formation to better enable landscape-led climate actions within projects.

> I have strategically involved myself in more of those strategies and policies that are looking to get planning changes because of some of the shortfalls in our briefs or our capacity to do what we know we should be doing from a climate and biodiversity perspective.

Working between disciplinary boundaries was seen as a key activity and area of successful boundary work for the practice, as was developing a reputation for effectiveness in this area of contribution. However, the need to get involved in early stages of projects to ensure impact is not always met.

> A lot of our work is associated with multidisciplinary teams and especially architects. On average, big projects have 20 consultants and that includes sustainability consultants and engineers.
> I think increasingly, there is recognition for how we do think about those things, and the importance of us being engaged early doesn't necessarily always reflect in our fee to then be commensurate with time.

Key Actions and Challenges

The practice leader indicated that the parts of the life stages they worked across most often were between strategic planning and design to renewal and recovery, and less often being involved at the change-initiation phase. Planning for renewal and recovery in projects is currently of interest and beginning to become part of practice operations.

A clear challenge the interviewee identified concerned the ability for the practice to "prioritize innovation" as a climate action within a constrained economic context. This was contextualized in relation to rising construction costs influencing value management within projects. An additional challenge is recognizing boundary spanning work in project fees.

> Due to fiscal constraints, the level of ambition and all the policy around what people need to achieve through their different procurement models and in acquitting all these things is not necessarily commensurate with the budget provision.

Another of the practice leader's observations was that the phases of a project lifecycle created temporal boundaries. Even within a single project these were not always aligned or discreet, meaning that at times the practice might be working across multiple project phases and actions.

This case highlights the importance of boundary spanners to enable the incremental uplift of climate research and policy into practice and across disciplines. This case study provides an example of boundary work that crosses the policy and practice divide in a recursive loop, influencing and forming policy through evidence-based knowledge, followed by implementation into practice that then informs policy. It also highlights the challenge of making this boundary work visible within organizations or single projects, and the need to be able to evaluate impact across a network of cascading projects and collaborations.

4.4.2 *Case Study 2: Not-For-Profit Collective, Dark Matter Labs, Denmark*

Boundary Roles

The interview with the co-director of the Copenhagen-based Dark Matter Labs collective explored the boundary spanning function within the context of climate action, sustainability transitions, and systemic change. The interviewee defined their role not as a direct innovator, but as a facilitator of systemic change, enabling collaboration across disciplines, sectors, and geographies. In their work, boundary spanning is considered crucial for bridging the gaps between diverse actors, including designers, developers, policymakers, and investors—groups that are typically disconnected but essential for tackling climate change. Identifying "missing links" in the climate-action ecosystem, bringing together the necessary actors, and ensuring that long-term relationships are established to drive transformative change were key elements in defining boundary roles. The interviewee noted that this involves creating the conditions for collective learning and action, pushing beyond the traditional roles of advisory or program design. They saw this boundary role as being often about enabling the change, rather than directly implementing it, by setting up a space where innovation can emerge through collaborative dialogue and shared visions:

> Maybe what we try to do is not find the blueprint for the future, like some do. We are trying to enable the change, enable the learning processes, enable innovation, not do the innovation ourselves.

Types of Boundaries

The interviewee asserted that a significant aspect of their boundary spanning work was navigating the various types of boundaries that affect the built environment. These boundaries are not only disciplinary, but also institutional, cultural, and spatial in nature. Navigating these boundaries includes taking actions to ensure the necessary conversations between stakeholders happen at the right stages, enabling a more holistic approach to sustainable urban development. The interviewee particularly emphasized crossing sectoral and disciplinary boundary types:

> ...to detect what are the actors you need, so what are things you need to [enable] change across the systems, what are the disciplines and then who are the actors that need to sit at the same table at the same time to enable the systemic shift.

Key Actions and Challenges

In mapping their boundary spanning actions across life stages, the interviewee noted that their work crosses many stages, beginning with climate risk assessments in urban areas at the change-initiation stage, as well as connecting actors across sectors to identify key climate vulnerabilities and adaptation needs in the project-consideration stage. They are also deeply involved in promoting "designing for circularity" within the built environment at the design and construction phase, advocating for maximizing the life cycles of built assets and reusing existing building stocks instead of new developments as a way to reduce emissions and resource extraction. The co-director's role also extends into policy change, particularly in the regulatory and financial domains (i.e., costing and approvals), where they work to ensure that policy frameworks align with the ambitious goals of decarbonization and circular economy practices, and that they facilitate a transition to a new normal, rather than only preventing unwanted outcomes.

Despite the interviewee's significant boundary spanning efforts, they often face several challenges that hinder the effectiveness and scale of climate action. They identified one of the primary barriers as the incomplete coalitions of actors involved in climate initiatives. Frequently, key stakeholders are absent from the dialog, preventing full system change. Despite aiming for a more integrated approach, the lack of alignment between these actors can lead to stagnation or slow progress, as decisions are made without the necessary support or understanding from all involved. Furthermore, the interviewee highlighted there are significant cultural barriers that impede the integration of innovation, including the use of innovative materials and technologies in built environments. Often, the existing systems—such as finance, insurance, and regulations—are not ready or flexible enough to support these changes, and these barriers slow the pace of climate action. As a result, the interviewee's work often involves not only connecting key actors but also challenging the systems and cultural norms that reinforce the status quo.

This example demonstrates that despite multiple challenges for enabling systemic change, remaining optimistic is a key attribute of boundary actors. It shows that the key to unlocking systemic change lies in fostering collaborative partnerships, simplifying complexity, and encouraging innovative thinking that can break through the entrenched barriers hindering progress, especially when investing in climate initiatives:

> We communicate with investors with pension funds, long-term investors, when we tell them the risks of today are not the risks of tomorrow. So in future, the main problem will be climate-adapted or resilient neighborhoods. It's not about building a new house that you will sell. It's more about climate-adapted neighborhoods. So that would be the long-term investment for the future risks.

4.4.3 *Case Study 3: Regional Local Government, Menjez Municipality, Lebanon*

Boundary Roles

The final case study focuses on a small regional municipality in the north of Lebanon, close to the border with Syria, representing a Global South case study. The selected interviewee is known as a climate-conscious leader and a champion for urban sustainability and climate solutions. Despite limited technical and financial resources, and the need to manage pressures from influxes of Syrian refugees, the municipality has become a regional sustainability and climate model, joining a global city network in 2014 and developing one of the first Sustainable Energy Action Plans (SEAP) in Lebanon.

The interview revealed the key role that the interviewee played as a connector, who often worked between academia and practice, in advancing climate actions. The way the interviewee framed their boundary spanning role revolved around their ability to bring together the right consortium of actors across universities, NGOs, and government bodies, to work collectively on climate-innovation initiatives. The interviewee emphasized the importance of understanding the world of academia, including its different "language" and ways of working, and the urge for innovative solutions and progressive thinking as key drivers for making collaborations with universities fruitful. They believed that their second role as a university teaching academic in their original field (allied medical and health), has helped them value the integration of science, and build the required skills and competencies in effectively communicating and building collaborations with researchers across universities in Lebanon and abroad.

> I come from [the] academic sector. And it's important for us in academic sector to work in exchange of good practices.

This collaboration was seen as critical in gaining success in securing multiple external funds, emphasizing the importance of co-developing evidenced-based knowledge through transdisciplinary consortiums led by universities.

Types of Boundaries

One of the main boundary types that the interviewee noted was, "trans-border boundaries." Most of the external funding that was secured for sustainability initiative came from successful application to "calls for proposals," which often have a key criterion of demonstrating trans-border collaborations. This was seen as a positive incentive to work across national boundaries, and build relationships with other regional municipalities, including in France and Germany. Crossing geographical boundaries also entailed working across socio-cultural boundaries, as well as knowledge-system boundaries. The interviewee particularly emphasized that working with universities required the establishment of trust among partners, including giving researchers the agency to lead, but also acknowledging the integrity of different knowledge systems (including communities' place-based knowledge and Indigenous knowledge).

Key Actions and Challenges

Given the role of local governments in working across different scales and sectors of the built environment, the interviewee stated that most of their actions crossed all life stages from change initiation to end of life of projects. One key action that they emphasized was building the right partnerships at project initiation.

Given the limited access to financial resources in most regional councils, a key challenge the interviewee highlighted was building trust with international funding bodies, which requires developing partnerships beyond national boundaries. Working closely with universities and NGOs was considered an important facilitating factor. The interviewee perceived the role of the council as an intermediary that enables allocation and funneling of funds to support researchers in developing evidence-based knowledge from practice:

> The funding agencies have more flexibility to give money to the university. And not to a municipality or the public sector... and to mobilize funds from funding agencies [to universities is critical].

The interviewee highlighted that funders often perceive universities as providing credibility, rigor, and independence, particularly in contexts where political stability, probity, and integrity are prominent issues.

A key identified opportunity in building effective relationships across boundaries was perceived as the role that global city networks can play in supporting member cities to set more-ambitious climate targets, and to identify new actions and partnerships to achieve them, highlighting the boundary role that networks play across national boundaries. This case also highlighted the important role that local governments can play as intermediaries between communities, politicians, practices, and universities (Stephens et al., 2024).

4.5 Discussion and Conclusions: Interactional Expertise, a Key to Effective Boundary Spanning

Table 4.2 synthesizes the key findings from the three case studies.

A synthesis of the three case studies indicates that boundary spanning roles for climate actions can be carried out in two forms: through individuals with strong ***interpersonal*** and ***communication attributes***, and through organizations as intermediaries and ***connectors for systemic change***. These organizations often absorb (or employ) actors who embody the skills and

Table 4.2 Findings from three case studies

Case Study	*Nature of practice*	*Types of boundaries crossed*	*Life-stage focus and boundary actions*	*Challenges and opportunities*
Case 1: Medium-sized private landscape and urban-design practice (Australia)	Recognized for multidisciplinary collaboration and a focus on urban regeneration, climate resilience, and community-centered design	Vertical (governance)	Life stages from strategic planning up to renewal and recovery. Sometimes, but less, often change initiation	Prioritizing innovation with constrained budgets Recognition of boundary spanning work in project briefs
Case 2: Large-scale not-for-profit design collective (Denmark)	Transdisciplinary not-for-profit focused on system-level design, innovation, and experimental civic infrastructure	Horizontal (disciplinary, sectoral)	Most life stages, particularly change initiation and end-of-life of projects Actions include policy change, aligning investment, and awareness-building in maximizing existing building utilization and circularity	Finding the right missing actors, such as policy-change agents or investors, and bringing them to the table, building trust, encouraging cultural-mindset shifts
Case 3: Small rural municipality (member of global city network) (Lebanon)	Known as a climate-model municipality and key leader; projects in biodiversity, water security, energy, and infrastructure	Spatial (geographical, transborder) Horizontal (science–practice)	Most life stages Actions include partnership formation at the project-consideration stage	Trust building in decentralized governance, dealing with issues related to political stability and probity

competencies of boundary spanners themselves, while having substantial expertise in climate-resilient cities. This is the case of frontrunner practices such as case study 2, which is at the forefront of reimagining infrastructure, finance, and governance models, and crafting bold frameworks that shift societies towards regenerative, resilient futures. These types of practices and organizations often act as *facilitators and enablers of innovation*, rather than offering specific solutions. In contrast, the local council case suggests that key individuals who are proactive, driven, and *well-connected with established networks* of alliances and partners are crucial in positioning local governments as boundary organizations. This resonates with the findings of Stephens et al. (2024) highlighting the important attributes of *leadership that is vision-driven* and capable of bridging divides and enabling collaboration. Similarly, small, private design practices with individuals incorporating boundary spanning activities can *influence practice operations*. These practices, and the individuals within them, expand from the provision of design services into advocacy roles that both enable the transfer of knowledge between designers, clients, and communities, and directly inform decision processes by implementing climate-responsive strategies through design.

The interviews further revealed that the *relational* dimension of boundary spanning proves critical in mobilizing climate action in cities, where establishing *effective networks and partnerships* across multiple life stages helps navigate existing challenges such as competing political interests, budget constraints, and entrenched systemic inertia. This aligns with the review by Neal et al. (2023), who found that *facilitating relationships* was the dominant strategy used by knowledge brokers, intermediaries, and boundary spanners, followed by *disseminating evidence*. A key facilitator for these strategies is *building trust*. Trust building, a critical part of boundary spanning, helps sustain access to the different knowledge systems and facilitates the transfer of knowledge and resources (Leck & Roberts, 2015). Applying the credibility of universities to mobilize resources towards climate initiatives emerged as an enabler for boundary work. Nevertheless, without trust, there is enormous inertia at the interfaces between knowledge systems, which hinders collective and inclusive actions. Boundary spanners help convene and create trust and shepherd new and old coalitions, including with communities and holders of local and Indigenous Knowledge (Gibbs & Ross, 2025); this requires strong interpersonal and communication skills, as well as time to build trusted relationships (Moosavi et al., 2025b).

Successful boundary spanners develop deep trust relationships across different types of boundaries, including institutional divides, allowing them to *facilitate difficult conversations* about climate trade-offs, future risks, and uncertainties (Bednarek et al., 2018; Meadow et al., 2015). They often possess what Williams (2002) describes as "interactional expertise": sufficient fluency in both scientific and policy discourses to create mutual understanding without necessarily possessing the full practical competence in either domain. This enables boundary spanners to translate between the probabilistic language of climate science and the decision frameworks of urban governance and investment, helping cities move from knowledge about climate risks to implementable adaptation and mitigation strategies. The work of organizations and individuals in connecting to investors and communicating future climate risks to create awareness of long-term investment needs is identified as an important enabler for supporting boundary work.

The findings from the three different examples further advance theories of boundary work by highlighting the need for differentiated support for its various forms. The different profiles of boundary spanners interviewed in this research suggest that various contexts give rise to different profiles of boundary actors with particular attributes. This highlights the need for an enhanced understanding of the different forms of boundary work needed in the context of climate actions in cities, pointing to an area for further research. The limitations of the small sample size used in this study could be addressed by incorporating complementary methods such as network-knowledge analyses (Posner & Cvitanovic, 2019) and investigating additional case studies, to help further evaluate and monitor the impacts of boundary spanners in climate decisions, as well as describing patterns of knowledge exchange among decision-makers.

Finally, boundary spanning needs to be embedded in broader governance reforms to help build capacity and inform sustained investment in innovative climate actions. A greater recognition of their contribution and impact is needed (Palmer et al., 2016); boundary organizations such as the Stockholm Resilience Centre have created spaces where boundary spanning actors receive explicit recognition and support for translational work. Events and platforms such as the Innovate4Cities conference create a safe space for connecting boundary actors and providing opportunities for them to engage in knowledge exchange, but also for enhancing recognition of the key role they play in climate actions.

This chapter offers an expansive, yet nuanced, picture of boundary spanning in its different practiced forms as lived experiences, and highlights the need for going beyond technical innovation, towards social innovation in relational dimensions, and towards interactional expertise for enabling systemic change. Establishing a shared understanding of different practiced forms of boundary spanning further contributes to the development of a professional community of practice and an enhanced recognition of their impact. Boundary spanning should be perceived not just as a theoretical ideal, but as an operational necessity that embraces relational and multi-actor collaborations across knowledge boundaries, to co-create climate-responsive strategies for cities.

Acknowledgements The authors would like to thank the interviewees who generously shared their experiences and insights. Special thanks to Ivana Stancic, Georges Yussef, Claire Martin, and Sarah Bell as well as the organizers of Innovate4Cities for enabling the initial dialogues that formed this research project.

Competing Interests
The authors have no conflicts of interest to declare that are relevant to the content of this chapter.

References

Bednarek, A. T., Wyborn, C., Cvitanovic, C., Meyer, R., Colvin, R. M., Addison, P. F. E., Close, S. L., Curran, K., Farooque, M., Goldman, E., Hart, D., Mannix, H., McGreavy, B., Parris, A., Posner, S., Robinson, C., Ryan, M., & Leith, P. (2018). Boundary spanning at the science–policy interface: The practitioners' perspectives. *Sustainability Science, 13*(4), 1175–1183. https://doi.org/10.1007/s11625-018-0550-9

Burbach, M. E., Eaton, W. M., & Delozier, J. L. (2023). Boundary spanning in the context of stakeholder engagement in collaborative water management. *Socio-Ecological Practice Research, 5*(1), 79–92. https://doi.org/10.1007/s42532-023-00138-w

Carlile, P. R. (2004). Transferring, translating, and transforming: An integrative framework for managing knowledge across boundaries. *Organization Science, 15*(5), 555–568. https://doi.org/10.1287/orsc.1040.0094

Clark, W. C., Tomich, T. P., van Noordwijk, M., Dickson, N. M., Catacutan, D., Guston, D., & McNie, E. C. (2010). Toward a general theory of boundary work: Insights from the CGIAR's natural resource management programs.

Cross, R., Ernst, C., & Pasmore, B. (2013). A bridge too far? How boundary spanning networks drive organizational change and effectiveness. *Organizational Dynamics, 42*(2), 81–91. https://doi.org/10.1016/j.orgdyn.2013.03.001

Dąbrowski, M. (2017). Boundary spanning for governance of climate change adaptation in cities: Insights from a Dutch urban region. *Environment and Planning c: Politics and Space, 36*(5), 837–855. https://doi.org/10.1177/2399654417725077

Delozier, J. L., & Burbach, M. E. (2021). Boundary spanning: Its role in trust development between stakeholders in integrated water resource management. *Current Research in Environmental Sustainability, 3*, Article 100027. https://doi.org/10.1016/j.crsust.2021.100027

Edwards, A. (2011). Building common knowledge at the boundaries between professional practices: Relational agency and relational expertise in systems of distributed expertise. *International Journal of Educational Research, 50*(1), 33–39.

Eriksen, S. H., Simpson, N. P., Glavovic, B., Ley, D., Carr, E. R., Fernández-Carril, L., Hayward, B., Morecroft, M. D., Pathak, M., Pereira, J. J., Pörtner, H.-O., Roberts, D. C., Ruane, A. C., Sánchez-Rodríguez, R., Schipper, E. L. F., Solecki, W., Stringer, L. C., Totin, E., Birt, A., & Degvold, F. (2024). Pathways for urgent action towards climate resilient development. *Nature Climate Change*. https://doi.org/10.1038/s41558-024-02190-0

Fischer, L.-B., & Newig, J. (2016). Importance of actors and agency in sustainability transitions: A systematic exploration of the literature. *Sustainability,8*(5), 476. Retrieved September 12, 2025 from https://www.mdpi.com/2071-1050/8/5/476

Gibbs, M. C., & Ross, P. M. (2025). Boundary spanners: An Australian first nations perspective. *People and Nature, 7*(8), 1789–1795. https://doi.org/10.1002/pan3.70095

Goodrich, K. A., Sjostrom, K. D., Vaughan, C., Nichols, L., Bednarek, A., & Lemos, M. C. (2020). Who are boundary spanners and how can we support them in making knowledge more actionable in sustainability fields? *Current Opinion in Environmental Sustainability, 42*, 45–51. https://doi.org/10.1016/j.cosust.2020.01.001

Hatch, M. B. A., Parrish, J. K., Heppell, S. S., Augustine, S., Campbell, L., Divine, L. M., Donatuto, J., Groesbeck, A. S., & Smith, N. F. (2023). Boundary spanners: A critical role for enduring collaborations between Indigenous communities and mainstream scientists. *Ecology and Society, 28*(1), Article 41. https://doi.org/10.5751/ES-13887-280141

Hawkins, M. A., Rezazade, M., & Mohammad, H. (2012). Knowledge boundary spanning process: Synthesizing four spanning mechanisms. *Management Decision, 50*(10), 1800–1815. https://doi.org/10.1108/00251741211279611

Hurlimann, A., Moosavi, S., Bush, J., March, A., Prideaux, F., Warren Myers, G. (2025). Integrating climate change action across the built environment: A guide for transformative action https://doi.org/10.26188/27366297.v1

Hürlimann, A., Warren-Myers, G., Nielsen, J., Moosavi, S., Bush, J., & March, A. (2021). Towards the transformation of cities: A built environment process map to identify the role of key sectors and actors in producing the built environment across life stages. *Cities, 103454*. https://doi.org/10.1016/j.cities.2021.103454

Hürlimann, A. C., Nielsen, J., Moosavi, S., Bush, J., Warren-Myers, G., & March, A. (2022). Climate change preparedness across sectors of the built environment—A review of literature. *Environmental Science & Policy, 128*, 277–289. https://doi.org/10.1016/j.envsci.2021.11.021

IPCC. (2022). Climate change 2022: Mitigation of climate change. In *Contribution of working group III to the sixth assessment report of the intergovernmental panel on climate change*. Retrieved September 12, 2025 from https://www.ipcc.ch/report/ar6/wg2/

Karcher, D. B., Cvitanovic, C., Colvin, R. M., van Putten, I., Pozza, R. D., & Posner, S. (2025). Key attributes for effective knowledge brokering at the interface of environmental science and management. *Sustainability Science, 20*(1), 117–133. https://doi.org/10.1007/s11625-024-01575-6

Leck, H., & Roberts, D. (2015). What lies beneath: Understanding the invisible aspects of municipal climate change governance. *Current Opinion in Environmental Sustainability, 13*, 61–67. https://doi.org/10.1016/j.cosust.2015.02.004

Meadow, A. M., Ferguson, D. B., Guido, Z., Horangic, A., Owen, G., & Wall, T. (2015). Moving toward the deliberate coproduction of climate science knowledge. *Weather, Climate, and Society, 7*(2), 179–191. https://doi.org/10.1175/WCAS-D-14-00050.1

Moosavi, S., Hurlimann, A., Neilsen, J., March, A., Bush, J., & Warren-Myers, G. (2025a). Towards transformative climate actions in landscape architecture. *Landscape Ecology, 40*(120). https://doi.org/10.1007/s10980-025-02129-6

Moosavi, S., Daniela, P., & Stephan, A. (2025b). Metabolic thinking in planning and designing urban landscapes: Practitioners' perspectives on agency. *European Planning Studies*, 1–24. https://doi.org/10.1080/09654313.2025.2501565

Neal, J. W., Neal, Z. P., & Brutzman, B. (2022). Defining brokers, intermediaries, and boundary spanners: A systematic review. *Evidence & Policy, 18*(1), 7–24. https://doi.org/10.1332/174426420x16083745764324

Neal, J. W., Posner, S., & Brutzman, B. (2023). Understanding brokers, intermediaries, and boundary spanners: A multi-sectoral review of strategies, skills, and outcomes. *Evidence & Policy, 19*(1), 95–115. https://doi.org/10.1332/174426421x16328416007542

Palmer, M. A., Kramer, J. G., Boyd, J., & Hawthorne, D. (2016). Practices for facilitating interdisciplinary synthetic research: The National Socio-Environmental Synthesis Center (SESYNC). *Current Opinion in Environmental Sustainability, 19*, 111–122. https://doi.org/10.1016/j.cosust.2016.01.002

Posner, S. M., & Cvitanovic, C. (2019). Evaluating the impacts of boundary-spanning activities at the interface of environmental science and policy: A review of progress and future research needs. *Environmental Science & Policy, 92*, 141–151. https://doi.org/10.1016/j.envsci.2018.11.006

Satheesh, S. A., Verweij, S., van Meerkerk, I., Busscher, T., & Arts, J. (2023). The impact of boundary spanning by public managers on collaboration and infrastructure project performance. *Public Performance & Management Review, 46*(2), 418–444. https://doi.org/10.1080/15309576.2022.2137212

Stephens, W., van Steden, R., & Schoonmade, L. (2024). Boundary spanning in local governance: A scoping review. *Administration & Society, 56*(2), 99–144. https://doi.org/10.1177/00953997231219262

Taylor, A., Pretorius, L., McClure, A., Iipinge, K. N., Mwalukanga, B., & Mamombe, R. (2021). Embedded researchers as transdisciplinary boundary spanners strengthening urban climate resilience. *Environmental Science & Policy, 126*, 204–212. https://doi.org/10.1016/j.envsci.2021.10.002

Tribbia, J., & Moser, S. C. (2008). More than information: What coastal managers need to plan for climate change. *Environmental Science & Policy, 11*(4), 315–328. https://doi.org/10.1016/j.envsci.2008.01.003

van den Brink, M., Edelenbos, J., van den Brink, A., Verweij, S., van Etteger, R., & Busscher, T. (2019). To draw or to cross the line? The landscape architect as boundary spanner in Dutch river management. *Landscape and Urban Planning, 186*, 13–23. https://doi.org/10.1016/j.landurbplan.2019.02.018

van den Brink, M., van den Brink, A., & Bruns, D. (2022). Boundary thinking in landscape architecture and boundary-spanning roles of landscape architects. *Landscape Research*, 1–13. https://doi.org/10.1080/01426397.2022.2091121

van Dijk, T., Kempenaar, A., van den Brink, M., & Laeni, N. (2023). Boundary spanning in design-led strategic spatial planning: Lessons from post-Sandy rebuilding efforts. *International Planning Studies, 28*(3–4), 367–383. https://doi.org/10.1080/13563475.2023.2251703

Williams, P. (2002). The competent boundary spanner. *Public Administration, 80*(1), 103–124. https://doi.org/10.1111/1467-9299.00296

CHAPTER 5

The Role and Impact of City and Science Networks and Science for Urban Climate Solutions

Maria Dombrov, Daniel Pejic, Benjamin Leffel, Pablo Mariani, Giulia Melica, David Miller, Valentina Palermo, Cynthia Rosenzweig, Susannah Smetana Kagan, William Solecki, and Maryke van Staden

5.1 City and Science Networks: Functions Across Science–Policy–Practice

City and science networks (CSNs) are defined as knowledge generators, translators, and connectors within their communities of practice and between the science, policy, and practice divide. These networks are leading facilitators of science-based urban climate solutions by encouraging *interconnectedness*, *common objectives*, and *information*

M. Dombrov (✉) · C. Rosenzweig · S. S. Kagan · W. Solecki
Urban Climate Change Research Network, Columbia University Climate School, New York City, NY, USA
e-mail: m.dombrov@columbia.edu

C. Rosenzweig
e-mail: crr2@columbia.edu

C. Oke et al. (eds.), *A Global Research and Action Agenda on Cities and Climate Change Science*,
https://doi.org/10.1007/978-981-95-6517-7_5

sharing. Through science and collaboration, CSNs, as well as climate-focused networks, possess a well-versed understanding of complex and compounding climate change issues and work alongside cities to help lead the implementation of urban climate solutions.

S. S. Kagan
e-mail: ss5468@columbia.edu

W. Solecki
e-mail: wsolecki@hunter.cuny.edu

M. Dombrov · C. Rosenzweig
NASA Goddard Institute for Space Studies, New York City, NY, USA

D. Pejic
Melbourne Centre for Cities, University of Melbourne, Melbourne, Australia
e-mail: daniel.pejic@unimelb.edu.au

B. Leffel
University of Nevada Las Vegas, Las Vegas, NV, USA
e-mail: benjamin.leffel@unlv.edu

P. Mariani
United Cities and Local Governments, Brussels, Belgium
e-mail: p.mariani@uclg.org

G. Melica · V. Palermo
European Commission, Joint Research Centre (JRC), Ispra, Italy
e-mail: Giulia.MELICA@ec.europa.eu

V. Palermo
e-mail: Valentina.PALERMO@ec.europa.eu

D. Miller
C40 Centre for City Climate Policy and Economy, Toronto, ON, Canada
e-mail: dmiller@c40.org

W. Solecki
Department of Geography and Environmental Science, Urban Climate Change Research Network, Hunter College – CUNY, New York City, NY, USA

M. van Staden
carbonn Climate Center, ICLEI Local Governments for Sustainability, Bonn, Germany
e-mail: maryke.van.staden@iclei.org

CSNs operate at various scales, including internationally and regionally as well as at country, state, and municipal levels.[1] While there are few precise figures, recent research has identified at least 60 city networks globally working on climate change issues, many with global remits and large member bases (Cortes et al., 2022). CSNs are critical for both Global South cities to integrate climate mitigation and adaptation practices into sustainable development, and for Global North cities to reduce greenhouse gas emissions while systemically strengthening resilience.

To foster *interconnectedness*, CSNs interact directly within and across cities, supporting city leaders and policymakers in their decision-making processes. Membership-based CSNs, such as the C40 Cities Climate Leadership Group (C40), bring together nearly 100 international mayors to advance climate actions in their cities, with annual requirements to meet performance-based standards. Other CSNs, such as United Cities and Local Governments, facilitate peer-to-peer learning that brings together local and regional governments for dialogs on urban-resilience strategies and planning (Resilient & Inclusive Cities, 2025). CSNs simultaneously structure their own organizations with experienced climate leaders, practitioners, scientists, researchers, policy advisors, and project managers.

Common objectives of CSNs in cities include, but are not limited to, joint advocacy on climate issues, greenhouse gas emissions reductions, clean energy and technological advancement, climate-risk reduction through effective adaptation, and equitable climate-resilient development. CSNs provide various technical and knowledge-related resources to reduce greenhouse gas emissions, including emissions-inventorying software, access to climate policy templates, access to climate finance, group discounts on retrofits, and other tools critical for implementation (Abbott, 2018; Betsill & Bulkeley, 2004; Bulkeley et al., 2009). The Global Covenant of Mayors for Climate and Energy (GCoM), for example, offers the Global Common Reporting Framework, which helps cities to keep track of their greenhouse gas emissions-data and climate action reporting

[1] The term "city and science networks" (CSNs) are used to broadly encompass a range of organizations working to advance climate change actions across science–policy–practice sectors within and across cities. Other networks exist, such as "city networks", "membership networks", "transnational municipal networks", and "transnational climate networks" (Acuto & Leffel, 2021; Lee & Jung, 2018).

(GCoM, 2023).While CSNs are often aligned on overarching ambitions, each has its own sector expertise and thematic elements.

To encourage *information sharing*, CSNs are continuously developing and disseminating scientific information in various formats and forums through climate assessments, journal articles, books and reports, workshops, and webinars. CSNs, such as the Urban Climate Change Research Network (UCCRN), produce Assessment Reports on Climate Change and Cities (ARC3) that include city-level climate projections of temperature, precipitation, and sea level rise, as well as actionable policy recommendations (Rosenzweig et al., 2018).

CSNs are central to creating localized climate knowledge, but they also have an obligation to ensure that the information is usable, accessible, and applicable for a variety of city typologies in the Global South and North with varied governance mechanisms and financial resources. CSNs can help alleviate the implementation burden on cities by providing technical expertise and context-specific guidance on how best to reduce greenhouse gas emissions and adapt to the most pressing climate extremes. The uptake of this knowledge in cities tends to be gradual through pilot projects, and the transition from planning into implementation requires consistent effort and feedback loops. By incorporating monitoring, evaluating, and learning into implementation processes, CSNs can support cities in meeting ambitious long-term goals while encouraging open-access data and information.

5.1.1 Enabling Pathways for CSNs in Cities

In an increasingly urbanizing world, cities and local authorities are key players in the pursuit of global sustainability (Bai et al., 2019; Nagendra et al., 2018). The necessary societal transformation to achieve sustainable development goals, however, requires pooling resources and expertise from various actors, including government, universities, companies, and civil society (Trencher et al., 2014). For more than a decade, partnerships for urban sustainability have been recognized as key mechanisms for linking global-level research and knowledge to implementation efforts at the local or regional scale (Bai et al., 2019; Trencher et al., 2014).The number of CSNs at the interface between urban research, policy, and practice has grown dramatically, although with different focus areas and actors involved (Acuto & Rayner, 2016; Adams, 2012; Bai et al., 2019). Urban networks have emerged as transformative forces, reshaping urban governance, knowledge creation, and action by promoting dynamic, interconnected approaches (Glasbergen, 2010). CSNs enable collaboration and co-creation of solutions, scaling impacts across their reach and amplifying the effectiveness of urban initiatives (Muñoz-Erickson et al., 2017). Cities benefit from network participation through improved urban performance, with outcomes directly linked to the level of connectivity and active engagement within the network (Bai et al., 2019).

Cities addressing the climate change challenge can count on numerous science–policy—practice networks (Salvia et al., 2023; Ulpiani et al., 2025). Ambitious cities, such as those that expressed interest in joining the European Climate-Neutral and Smart Cities Mission (72.6% of 358 cities surveyed), demonstrated a strong engagement in various forms of inter-city collaborations, with city networks and initiatives as the most common type (Ulpiani et al., 2025). Networking appears to be the most important dimension of cities' engagement in the transition to climate neutrality (Salvia et al., 2023). Cities have regularly indicated that climate change is the top priority for their international engagement, and that city networking is one of their most effective international activities (Pejic et al., 2025). Involvement in national or transnational initiatives enables knowledge sharing, peer learning, and technical exchange. Cities can also count on tools and methodologies that are co-created and made accessible through these networks. Inter-city collaboration that materializes through joint projects, such as those funded under research framework programs, allows innovative ideas and approaches to be tested with the involvement of academia, the research community, NGOs, and the private sector (Trencher et al., 2014; Ulpiani et al., 2025).

5.1.2 CSNs and City–National–Global Connections

While CSNs are distinct in their foci, methods, and tools, they are unified through scientific assessments, such as the Intergovernmental Panel on Climate Change's Sixth Assessment Report (AR6), global treaties and frameworks like the UN Conference of the Parties Paris Agreement and UN Sustainable Development Goals (SDGs), and research agendas like the Global Research and Action Agenda (GRAA), developed jointly by CSNs and the IPCC. To align country-level ambitions with city-level actions, CSNs also use the Nationally Determined Contributions and National Adaptation Plans for the specific countries in which they work.

In addition to this guiding scientific information and these shared principles, CSNs develop their own roadmaps for climate planning, implementation, and the transfer of knowledge into practice. CSNs often assist cities in adopting their first Climate Action Plan (CAP), a foundational step in urban climate governance, especially as cities that adopt such plans—particularly in Europe—tend to meet their planning targets (Hsu et al. 2020; Sharp et al., 2011). One example is the DK2020 project, which called on all 98 municipalities in Denmark to develop a CAP that includes climate mitigation and adaptation for their district, and aligns with the C40 Cities Climate Action Planning framework (Damsø & Francart, 2024). All Danish municipalities were successful in developing CAPs with net-zero emissions targets by 2050, and rainfall and sea level-related hazards were prioritized in adaptation actions. This achievement was credited to high political buy-in, multi-tiered technical assistance, data and resource availability, and effective partnership structures and functions. DK2020 can serve as an informative framework for other countries with similar ambitions to unify and standardize their cities' climate agendas using CAPs.

5.1.3 Barriers and Limitations

While city networks play a crucial role in facilitating the exchange of knowledge, expertise, and resources to address urban climate change, they face several challenges that may undermine their effectiveness. One of the primary challenges is the growing complexity and fragmentation of these networks, which can lead to a lack of capacity to acquire new cities and engage all members effectively (Taveirne & Derudder, 2024). In addition, the proliferation of networks complicates coordination for city officials (Acuto & Rayner, 2016; Taveirne & Derudder, 2024), and the way cities shape their climate action is conditioned by the design and framework of their network membership (Schrage & Shrestha, 2025). Despite their contribution to knowledge generation and global policy processes, city networks require long-term financial and institutional support and continued engagement of membership in co-design and co-production activities (Bai et al., 2018, 2019).

A further challenge arises from the differences among members that manifest as difficulties in exchanging knowledge and best practices across contexts and capacities, particularly between frontrunner and follower cities, ultimately hindering the achievement of long-term sustainability. This can be further exacerbated by imbalances and inequalities across regions, particularly the North–South divide, in membership activities and actions (Bouteligier, 2013; Bulkeley et al., 2018; Schrage & Shrestha, 2025). Despite these limitations, city networks remain crucial for advancing urban climate solutions at scale and speed. Addressing the identified barriers can strengthen their role, ultimately accelerating the translation of scientific research into policy, practice, and implementation (Bai et al., 2019).

5.2 CSNs Engaging in Science-Driven Solutions

CSNs use the latest science to understand and stay informed on the anthropogenically induced changes to Earth, as seven of the nine planetary boundaries have already been crossed (Richardson et al., 2023; Sakschewski et al., 2025).[2] CSNs use the latest research to gain a grasp of the pace at which the global temperature is increasing and how it affects human and urban systems. CSNs can also use greenhouse gas emissions data to assess and monitor cities and countries' representative concentration pathways. Ultimately, through the presence and availability of peer-reviewed knowledge, CSNs can interpret this information to educate a variety of audiences and offer science-driven recommendations. Figure 5.1 shows the multitude of climate mitigation and adaptation tools with which CSNs work across regional, city, and district scales. CSNs also provide platforms and pathways to access the best available international data and practice, ensuring that domestic multi-level action is supported by proven and scalable solutions.

[2] "Anthropogenic" is defined as of, relating to, or resulting from the influence of human beings on nature (Merriam Webster, 2025), and "planetary boundaries" are defined as "the safe operating space for humanity with respect to the Earth system" (Rockström et al., 2009).

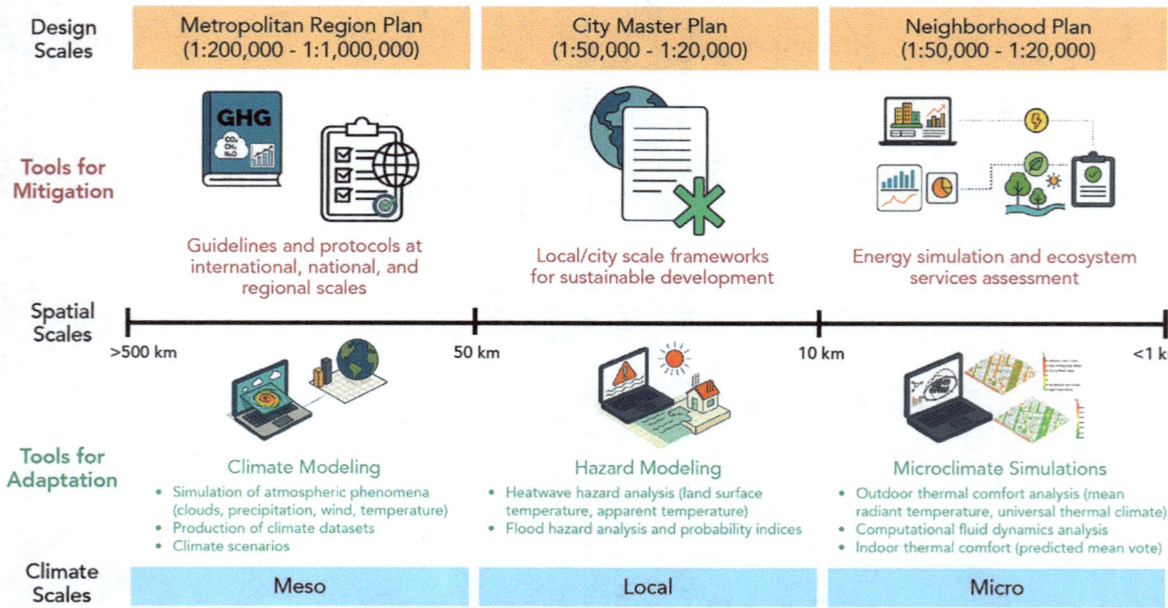

Fig. 5.1 Design, spatial, and climate scales of analytical and modeling tools for mitigation and adaptation (*Source* Raven et al., 2025, used with permission)

5.2.1 CSNs and the IPCC AR6 Summary for Urban Policymakers

Box 1.

The *Summary for Urban Policymakers of the IPCC Sixth Assessment Report (SUP)* is an example of the crucial role that city and science networks can play in bringing together academia and urban practitioners and bridging the science–policy divide (Revi et al., 2022). The IPCC's Sixth Assessment Cycle (AR6) highlighted the critical importance of cities and urban areas to the simultaneous aims of mitigation and adaptation.

Given the centrality of cities to climate crisis policy responses across both these domains, and the relative inaccessibility to urban policymakers of the more than 8,000 pages of analysis in AR6, a collaborative initiative was launched to develop a *Summary for Urban Policymakers.* This effort, supported by GCoM, Indian Institute for Human Settlements, Resilience First Business Network, Resilience Rising, and the German Government brought together more than 40 IPCC authors to produce three summary reports of the IPCC AR6 working groups, focused on: 1) what the latest physical science of climate change means for cities and urban areas; 2) what the latest research on impacts, adaptation, and vulnerability means for cities and urban areas; and 3) what the latest science on climate change mitigation means for cities and urban areas.

Among the detailed analysis, some key findings indicate:

- Even with strong emissions reduction, many cities and urban areas will experience more frequent floods, droughts, extreme heat, storms, and cyclones.
- By 2100, between 158 to 510 million urban residents will be affected by coastal flooding due to sea level rise.
- Climate change impacts will disproportionately impact marginalized urban communities with outsized risks for smaller to medium-sized cities in Asia, Africa, and Central and South America.
- All types of cities need urgent transformation to adapt to a changing climate (Klaus, 2023).

Rather than solely summarizing key findings for urban policymakers, this process also convened city and business leaders globally through regional fora to gain their input on the summary reports and produce an *Action Agenda* (Klaus et al., 2022). Both the SUP and Action Agenda were launched at COP27 in Sharm-El-Shaikh, Egypt.

The SUP process was made possible both by the financial support of key partners, such as the German government and, critically, by the established city and industry networks which could facilitate regional convenings of key actors and function as interfaces between IPCC authors and implementation practitioners. This case reveals how networks at the border of the science–policy nexus can complete the daunting task of translating thousands of pages of interdisciplinary scientific evidence into practical guides for implementation at the city-scale. While there is currently no systematic evidence of SUP adoption, many cities reported through their network collaborations that the outputs were valuable for their climate change teams, and national supporters, such as the German government, actively helped disseminate the reports to their municipalities.

The urban focus of AR6 and concentrated advocacy from urban supporters and key city networks have resulted in a *Special Report on Climate Change and Cities* to be included as part of AR7. This report, which will be the first delivered to the IPCC in the AR7 reporting cycle, is being developed by 97 urban climate experts, many of whom work closely with urban practitioners as part of their research. This special report should produce nuanced findings of more direct applicability for up-to-date responses to the climate crisis by urban practitioners.

5.3 Contributions to Climate Policy and Governance

Many CSNs connect climate science to evidence-based policy at regional and local levels; this is often referred to as the science–policy nexus.[3] Engaging city networks in the science–policy nexus helps to accelerate climate solutions and achieve outcomes, such as the co-creation of research-action agendas, the production of science-based knowledge or assessments using a transdisciplinary approach, and the creation and implementation of mechanisms for disseminating and translating such knowledge (Acuto et al., 2024; Heikkinen et al., 2020; Solecki et al.,

[3] "Science–policy nexus" refers to a collaborative public policy approach that makes use of the co-production of knowledge between climate experts, practitioners, and policymakers to achieve evidence-based, actionable policy (Bradaschia et al., 2024; Hunter et al., 2022).

2021). This section highlights outcomes of the science–policy nexus and identifies the role that city networks play in achieving them.

5.3.1 Science–Policy Nexus Outcomes

Convening academics, policymakers, and practitioners to develop a shared research and action agenda is one of the ways in which city networks directly engage in the science–policy nexus. City-government networks, such as GCoMand Local Governments for Sustainability (ICLEI), are integral in bringing together people working on urban climate change issues across the science, policy, and practitioner communities at both the local and national level, which are often isolated from one another (Acuto et al., 2021, 2024). The GRAA is a significant outcome of the science–policy nexus in identifying gaps, setting a research agenda, and stimulating knowledge generation for the AR7 Special Report on Climate Change and Cities (Oke et al., 2022; Solecki et al., 2021). The GRAA represents the crucial role that city networks have in convening across stakeholder groups with a focus on research and innovation, to achieve outcomes that drive climate action within the science–policy nexus (Oke et al., 2022; Global Covenant of Mayors et al., 2024).

For cities to most effectively address climate change, grounding policy in knowledge generated through transdisciplinary or multi-disciplinary frameworks allows for innovation, multiple perspectives, and more inclusive outcomes (Hunter et al., 2022; Solecki et al., 2021). A transdisciplinary research approach refers to the production of scientific knowledge in collaboration with various sources engaging across disciplines (Hunter et al., 2022). The need to address multi-risk climate impacts requires using this interdisciplinary and collaborative approach not only to connect science, practice, and policy, but also to connect across scientific disciplines to generate innovative climate solutions (Bradaschia et al., 2024; Frantzeskaki et al., 2019). The complexity of climate impacts in cities requires the integration of knowledge across disciplines such as climate science, social science, and spatial planning expertise (Bradaschia et al., 2024).

Networks that connect climate specialists across disciplines at the city, regional, national, and global scales help facilitate collaborative research to further climate policy and action (Solecki et al., 2021). Collaborative urban climate policy within the science–policy framework requires the integration of the climate science research community with practitioners and policymakers to gain understanding of practical experience and local needs (Bradaschia et al., 2024). Using CSNs to encourage researchers, policymakers, and practitioners to collaboratively generate climate solutions accelerates opportunities for innovative and informed climate action in cities such as Durban and Rio de Janeiro (Raven et al., 2025). This form of interconnectedness enables learning from cities in both the Global South and Global North.

City networks serve a key role in disseminating and translating climate science and evidence-based solutions to policymakers and leaders, which is necessary for science to inform climate policies in urban areas. To help bridge the usability gap between science and policy, city and science networks can translate science for policymakers and disseminate knowledge through case studies and assessments of evidence-based climate solutions, such as the UCCRN City Solutions Case Study Atlas and C40 Knowledge Hub (Solecki et al., 2021; Rosenzweig et al., 2026).

Sharing evidence-based case studies of urban climate solutions is an effective way for city and science networks to facilitate peer-to-peer learning between cities (Lee & Jung, 2018). City policymakers and decision–makers can look to examples of climate actions being taken in other cities to provide guidance and direction for implementing similar interventions in their own municipalities (Lee & Jung, 2018; Solecki et al., 2021).

5.3.2 *Overcoming Obstacles and Opportunities at the Urban Scale*

With the increasing variability of future climate conditions and their multiple impacts on cities, there is an immediate need and opportunity to address the complex and interconnected challenges that arise (Bradaschia et al., 2024; Krueger et al., 2022). Climate impacts experienced by cities are more complex than many existing climate plans or infrastructure and urban systems are designed to address. These impacts are exacerbated by other compounding factors, such as economic volatility, energy demands, access to water and food, and informality (Krueger et al., 2022). Emerging cities, especially those in the Global South, have increasing needs in the face of rapid urbanization and development (Parnell, 2018). Figure 5.2 shows that cities in low latitudes are projected to experience a disproportionate increase in extreme heat under the SSP5-8.5 global-warming scenario between 2021 and 2040 (Reckien et al., 2025). In response to these growing risks, cities such as Dhaka, Lagos, and Nairobi increasingly participate in city networks to access technical guidance and peer learning that strengthen their capacity to manage stressors, including extreme heat, water scarcity, and rapid urbanization (Krueger et al., 2022; Parnell, 2018). Building on this support, CSNs can further ensure that adaptation and mitigation planning enable just transitions by recognizing the benefits and burdens of climate actions, increasing procedural justice of vulnerable groups, and including justice considerations in the monitoring and evaluation of adaptation.

Cities have a need for expertise across disciplines to respond to the increasing complexity of the impacts of climate change. The interconnectedness of climate impacts with stressed infrastructure and urban systems, sensitive ecological conditions, and other economic constraints require innovative and collaborative approaches that integrate knowledge across disciplines (Lobo et al., 2023). In addition to scientific and technical expertise, cities also need expert knowledge on economics and other social sciences to develop integrated strategies that address multiple and simultaneous climate impacts (Lobo et al., 2023). The demand for this integrated approach creates an opportunity for city networks to bring together expertise across disciplines and at multiple levels of government to address these multi-risk challenges, drawing on both policy and practice (Bradaschia et al., 2024; Solecki et al., 2021).

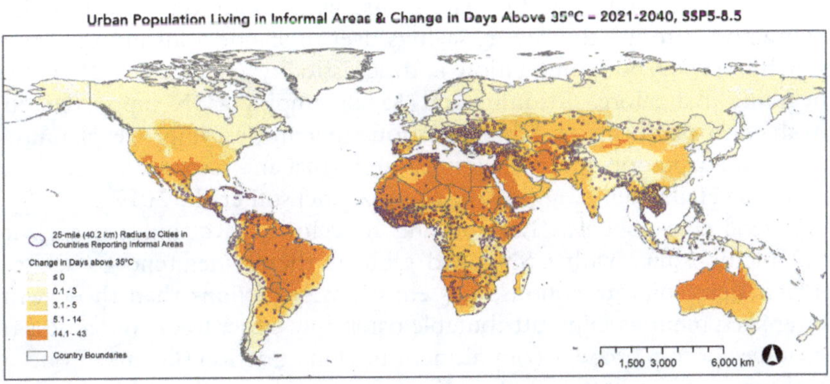

Fig. 5.2 UN-reported informal areas using a 2,525-mile radius. Change in number of extreme hot days in the near future (2021–2040) relative to 1995–2014 using scenario SSP5-8.5 (*Source* EM-DAT, 2023; Reckien et al., 2025, used with permission)

City networks can also play a role in expanding the available information and perspectives from emerging cities and underrepresented urban communities (Solecki et al., 2021). As CSNs support the dissemination and translation of knowledge, networks such as C40, GCoM, and UCCRN are further developing innovative platforms and technologies in response to the needs of city policymakers and practitioners to more easily share data, knowledge, and case studies, to facilitate peer learning. Asclimate impacts in cities are complex and interconnected with many other challenges and risks, CSNs have an important role to play in linking expertise across disciplines and various societal actors at different governance levels to facilitate transformative solutions.

5.4 Role of CSNs in Practice

CSNs have facilitated direct practice in urban planning and implementation by partnering with environmental consulting firms that specialize in building design, infrastructure, and energy systems. For instance, C40's partnership with ARUP provided cities with data-driven insights into the most productive and cost-effective policy actions (Gordon, 2020), while its collaboration with Honeywell through the Building Retrofit Program enabled member cities such as Melbourne to implement energy-performance contracts for government buildings (HBS, 2017). These contracts allow cities to reduce energy use with no upfront cost, repaying contractors through the energy savings over time—demonstrating a practical model for both mitigation and accessibility (Leffel, 2022). Cities including Bangalore, Kisumu, and Kigali employ CSN-supported co-production platforms to build emissions inventories, improve planning, and pilot adaptation actions tailored to informal and resource-constrained contexts (Heikkinen et al., 2020; Muñoz-Erickson et al., 2017).

Several cities, such as London and Barcelona, have multiple concurrent memberships with CSNs, and globally these cities tend to achieve greater per capita greenhouse gas emissions reductions than those with fewer such memberships, attributable ostensibly either to greater access to resources or to a broader commitment to climate action (Greater London Authority, 2025; Leffel, 2021). Korean cities that are ICLEI members have seen energy-consumption reductions ranging from 4.5% to over 20%, depending on their local and regional affiliations, underscoring the potential of CSN membership for energy-efficiency gains (Kim & Yun, 2024). Similarly, cities globally that are CSN members tend to reduce power-plant emissions more sharply, reflecting the cumulative effect of shared technical knowledge and policy diffusion (Grant et al., 2023). Cities such as San Francisco and Knoxville have made use of ICLEI's tools for emissions inventorying while also contracting local firms for tailored support, demonstrating how CSNs support implementation (Burns et al., 2005; City of San Francisco, 2004).

An emergent area of research examines interactions between climate-active businesses that adopt corporate sustainability practices and the climate-active cities in which they operate. Evidence shows that such cities often host companies with more aggressive emissions-reduction trajectories (Leffel et al., 2024). Recognizing this synergy, CDP and C40 launched the City-Business Climate Alliance to align urban planning and private-sector innovation at the local level (C40, CDP, & WBCSD, 2024). This partnership model further extends the reach of CSNs into energy access, technological upgrades, and financing arrangements for climate solutions. More research is needed in this area, as cross-sector impacts may offer more potential than is immediately evident. Recent studies indicate that firms located in C40 member cities have higher ESG (sustainability) scores due to climate-active C40 city governments, signaling the importance of corporate sustainability, which attract firms to collaborate and derive financial and other benefits (Wei & Wei, 2024). Further, mitigation impacts may be possible through these partnerships, as Kılkış et al. (2025) show, finding that co-located cities and businesses with existing net-zero and science-based targets could mutually elevate their emissions-reductions if they worked together.

5.5 Accelerating Science–Policy–Practice Interactions

To achieve emissions-reduction targets, adapt effectively to climate extremes, and reduce inequities, a coordinated effort among cities and CSNs is vital. Strengthening consistent, purposeful exchange between networks is especially important, with platforms such as the Innovate4Cities Conference providing opportunities to align priorities, clarify roles, and reduce impact duplication of effort. CSNs must also build on more than 30 years of advocacy that have established durable partnerships between networks and global environmental governance, ensuring that international frameworks remain targeted toward effective urban climate action. The IPCC *Special Report on Climate Change and Cities* represents a key milestone in this trajectory.

Within CSNs, there remains a need for more South–South collaboration and expanded opportunities for Global South leadership in governance and agenda-setting. These efforts should be paired with deliberate engagement of intermediary and smaller cities, which remain underrepresented in international environmental exchange. Recent analysis of environmental networks that span the Global North—South divide shows that while policy ideas often flow from North to South, cities adapt and reshape them to fit local contexts, generating their own innovations in the process.

Translating existing scientific data into actionable, digestible information should be prioritized. Developing open-access, user-friendly catalogs of ongoing and completed urban projects, paired with clear points of contact, will help cities to identify relevant models and accelerate implementation. Expanding assessments and case studies into additional languages will also enable broader participation. Open data on the global landscape of city networks further creates opportunities to understand how network activity shapes urban climate action and where gaps remain.

Despite persistent and substantial challenges, CSNs remain essential conduits for generating, translating, and sharing knowledge. Expanding their reach, resources, and geographic focus will be critical for enabling cities worldwide to accelerate mitigation and adaptation at the scale and speed required.

Acknowledgements The authors acknowledge Chiara Morfeo, of C40 Cities, for her contributions to the development of this chapter.

Competing Interests

The authors have no conflicts of interest to declare.

References

Adams, J. (2012). The rise of research networks. *Nature, 490*(7420), Article 7420. https://doi.org/10.1038/490335a

Abbott, K. W. (2018). Orchestration. In A. Jordan, D. Huitema, H. van Asselt, & J. Forster (Eds.), *Governing climate change: Polycentricity in action?*

(pp. 188—209). Cambridge University Press. https://doi.org/10.1017/9781108284646.012

Acuto, M., & Leffel, B. (2021). Understanding the global ecosystem of city networks. *Urban Studies, 58*, 1758–1774. https://doi.org/10.1177/0042098020929261

Acuto, M., Pejic, D., Mokhles, S., Leffel, B., Gordon, D., Martinez, R., Cortes, S., & Oke, C. (2024). What three decades of city networks tell us about city diplomacy's potential for climate action. *Nature Cities, 1*, 451–456. https://doi.org/10.1038/s44284-024-00081-1

Acuto, M., & Rayner, S. (2016). City networks: Breaking gridlocks or forging (new) lock-ins? *International Affairs*, *92*(5), Article 5. https://doi.org/10.1111/1468-2346.12700

Bai, X., Colbert, M., McPhearson, T., Roberts, D., Siri, J., Walsh, B., & Webb, B. (2019). Networking urban science, policy and practice for sustainability. *Current Opinion in Environmental Sustainability, 39*, 114–122. https://doi.org/10.1016/j.cosust.2019.08.002

Bai, X., Dawson, R. J., Ürge-Vorsatz, D., Delgado, G. C., Salisu Barau, A., Dhakal, S., Dodman, D., Leonardsen, L., Masson-Delmotte, V., Roberts, D. C., & Schultz, S. (2018). Six research priorities for cities and climate change. *Nature, 555*(7694), 23–25. https://doi.org/10.1038/d41586-018-02409-z

Betsill, M., & Bulkeley, H. (2004). Transnational networks and global environmental governance: The cities for climate protection program. *International Studies Quarterly, 48*, 471–493.

Bouteligier, S. (2013). Inequality in new global governance arrangements: The North–South divide in transnational municipal networks. *Innovation: The European Journal of Social Science Research, 26*(3), 251–267. https://doi.org/10.1080/13511610.2013.771890

Bulkeley, H., Schroeder, H., Janda, K., Zhao, J., Armstrong, A., Chu, S. Y., & Ghosh, S. (2009). Cities and climate change: The role of institutions, governance and urban planning. *Change, 28*, 30.

Bulkeley, H., Betsill, M., Compagnon, D., Hale, T., Hoffmann, M., Newell, P., & Peterson, M. (2018). Transnational governance: Charting new directions post-Paris. In *Governing climate change: Polycentricity in action?* (pp. 63–80). Cambridge University Press. https://doi.org/10.1017/9781108284646.005

Burns, T., Roberts, J., & Truesdale, J. (2005). *Public buildings energy retrofit: Knoxville case study.*

C40 Cities, CDP, & World Business Council for Sustainable Development (C40, CDP, & WBCSD). (2024). *City-business climate alliance.*

City of San Francisco. (2004). *Climate action plan: Local actions to reduce greenhouse gas emissions.*

Cortes, S., Van Der Heijden, J., Boas, I., & Bush, S. (2022). Unpacking the heterogeneity of climate city networks. *Cities, 121*, Article 103512. https://doi.org/10.1016/j.cities.2021.103512

Damsø, T., & Francart, N. (2024). *DK2020: A model for multi-level cooperation*. https://concito.dk/files/media/document/DK2020%20A%20model%20for%20multi-level%20cooperation.pdf. Accessed February 20, 2026.

EM-DAT: The International Disaster Database. (2023). www.emdat.be. Accessed June 12, 2025.

Frantzeskaki, N., McPhearson, T., Collier, M. J., Kendal, D., Bulkeley, H., Dumitru, A., & Pintér, L. (2019). Nature-based solutions for urban climate change adaptation: Linking science, policy, and practice communities for evidence-based decision-making. *Bio, 69*(6), 455–466. https://doi.org/10.1093/biosci/biz042

GaWC Dataset 29: Directed World City Network Centrality Scores for 12,802 Cities, 2016. https://gawc.lboro.ac.uk/gawc-worlds/gawc-data/dataset-29/. Accessed February 20, 2026.

GCoM. (2023). Common reporting framework. https://www.globalcovenantofmayors.org/wp-content/uploads/2023/11/CRF7-0-2023-09-14-final.pdf. Accessed February 20, 2026.

Glasbergen, P. (2010). Global action networks: Agents for collective action. *Global Environmental Change, 20*(1). https://doi.org/10.1016/j.gloenvcha.2009.09.002

Gordon, D. (2020). *Cities on the world stage: The politics of global urban climate action*. Cambridge University Press.

Granceri Bradaschia, M., Magni, F., & Musco, F. (2024). *Climate change adaptation, flood risk, and beyond: State of play in the science–policy-action nexus*. Springer Nature.

Grant, D., Leffel, B. & Johnson, E. (2023). Can transnational municipal networks mitigate the carbon pollution of the world's power plants? An empirical analysis. *NPJ Climate Action, 2*, 38.

Greater London Authority. (2025). *London-wide ultra low emission zone one year report*. London City Hall.

Honeywell Building Solutions (HBS). (2017). *Greener government buildings*.

Heikkinen, M., Karimo, A., Klein, J., Juhola, S., & Ylä-Anttila, T. (2020). Transnational municipal networks and climate change adaptation: A study of 377 cities. *Journal of Cleaner Production, 257*, Article 120474. https://doi.org/10.1016/j.jclepro.2020.120474

Hunter, N. B., Roberts, D. C., Sutherland, C., & Slotow, R. (2022). Co-creation between cities and climate change science achieves research and action agenda. *Current Research in Environmental Sustainability, 4*, Article 100189. https://doi.org/10.1016/j.crsust.2022.100189

Kim, M., & Yun, S.-J. (2024). The impact of transnational municipal networks on local energy consumption. *Urban Climate, 55*, Article 101861.

Kılkış, Ş, Bjørn, A., Bai, X., Liu, J., Whiteman, G., Crona, B., Andersen, L. S., Hasan, S., Vijay, V., & Sabag, O. (2025). City–company collaboration towards aligned science-based target setting. *Nature Sustainability, 8*(1), 54–65. https://doi.org/10.1038/s41893-024-01473-w

Klaus, I. (2023). *The remarkable scientific accomplishment of the latest IPCC Cycle*. Carnegie Endowment for International Peace. https://carnegieendowment.org/posts/2023/04/the-remarkable-scientific-accomplishment-of-the-latest-ipcc-cycle?lang=en. Accessed February 20, 2026.

Klaus, I., Eichel, A., Roberts, D., Schultz, S., & Revi, A. (2022). *Shared science, shared future—SUP action agenda*. Indian Institute for Human Settlements. https://doi.org/10.24943/SUPSV411.2022

Krueger, E. H., Constantino, S. M., Centeno, M. A., Elmqvist, T., Weber, E. U., & Levin, S. A. (2022). Governing sustainable transformations of urban social–ecological–technological systems. *NPJ Urban Sustainability, 2*(1), 10–12. https://doi.org/10.1038/s42949-022-00053-1

Lee, T., & Jung, H. Y. (2018). Mapping city-to-city networks for climate change action: Geographic bases, link modalities, functions, and activity. *Journal of Cleaner Production, 182*, 96–104. https://doi.org/10.1016/j.jclepro.2018.02.034

Leffel, B. (2021). Toward global urban climate mitigation: Linking national and polycentric systems of environmental change. *Sociology of Development, 8*, 111–137.

Leffel, B., Lyon, T. P., & Newell, J. P. (2024). Filling the climate governance gap: Do corporate decarbonization initiatives matter as much as state and local government policy? *Energy Research & Social Science, 109*, Article 103376.

Leffel, B. (2022). Climate consultants and complementarity: Local procurement, green industry and decarbonization in Australia, Singapore, and the United States. *Energy Research & Social Science, 88*, Article 102635. https://doi.org/10.1016/j.erss.2022.102635

Lobo, J., Aggarwal, R. M., Alberti, M., Allen-Dumas, M., Bettencourt, L. M. A., Boone, C., Brelsford, C., Broto, V. C., Eakin, H., Bagchi-Sen, S., Meerow, S., D'Cruz, C., Revi, A., Roberts, D. C., Smith, M. E., York, A., Lin, T., Bai, X., Solecki, W., … & Gauthier, N. (2023). Integration of urban science and urban climate adaptation research: Opportunities to advance climate action. *Npj Urban Sustainability, 3*(1), 32. https://doi.org/10.1038/s42949-023-00113-0

Muñoz-Erickson, T., Miller, C., & Miller, T. (2017). How cities think: Knowledge co-production for urban sustainability and resilience. *Forests, 8*(6), 203. https://doi.org/10.3390/f8060203

Nagendra, H., Bai, X., Brondizio, E. S., & Lwasa, S. (2018). The urban south and the predicament of global sustainability. *Nature Sustainability, 1*, 341–349. https://doi.org/10.1038/s41893-018-0101-5

Oke, C., Walsh, B., Assarkhaniki, Z., Jance, B., Deacon, A., & Lundberg, K. (2022). The city research and innovation agenda: Prioritizing knowledge gaps and policy processes to accelerate city climate action. *Journal of City Climate Policy and Economy*, *1*(1), 94–110. https://utppublishing.com/doi/full/10.3138/jccpe-2022.1.1.0007. Accessed February 20, 2026.

Parnell, S. (2018). Globalization and sustainable development: At the urban crossroad. *European Journal of Development Research, 30*(2), 169–171.

Pejic, D., Kling, S., Leavesley, A., Klaus, I., Trijsburg, I., Kosovac, A., & Acuto, M. (2025). *City diplomacy in response to multiple crises: The 2024 cities and international engagement survey*. Melbourne Centre for Cities, University of Melbourne. https://doi.org/10.26188/26866966.V1

Raven, J., Leone, M. F., Bhaduri, S., Braneon, C., Corbett, D., Driskell, D., Eicker, U., Fernandez, J. E., Jing, G., Hürlimann, A., Judah, I., & Viconti, C. (2025). Planning, urban design, and architecture for climate action. In W. Solecki, M. Pathak, M. Barata, A. Barau, M. Dombrov, & C. Rosenzweig (Eds.), *Climate change and cities: Third assessment report of the urban climate change research network*. Cambridge University Press. https://doi.org/10.1017/9781009643894

Reckien, D., Juhola, S., Haque, A. N., Khalil, H. A., Lemos, M. F., Lwasa, S., Niamir, L., Camilo Osorio, J., & Visconti, C. (2025). Justice for resilient development in climate-stressed cities. In W. Solecki, M. Pathak, M. Barata, A. Barau, M. Dombrov, & C. Rosenzweig (Eds.), *Climate change and cities: Third assessment report of the Urban Climate Change Research Network*. Cambridge University Press. https://doi.org/10.1017/9781009587105

Resilient and Inclusive Cities. (2025). *PLN40 resilient and inclusive cities*. https://learning.uclg.org/wp-content/uploads/2025/04/For-web-PLN40-Resilient_and_Inclusive_Cities.pdf. Accessed February 20, 2026.

Revi, A., Roberts, D., Klaus, I., Bazaz, A., Krishnaswamy, J., Singh, C., Eichel, A., Poonacha Kodira, P., Schultz, S., Adelekan, I., Babiker, M., Bertoldi, P., Cartwright, A., Chow, W., Colenbrander, S., Creutzig, F., Dawson, R., De Coninck, H., De Kleijne, K., … & Ürge-Vorsatz, D. (2022). *The summary for urban policymakers of the IPCC'S sixth assessment report*. Indian Institute for Human Settlements. https://doi.org/10.24943/SUPSV511.2022

Richardson, K., Steffen, W., Lucht, W., Bendtsen, J., Cornell, S. E., Donges, J. F., Drüke, M., Fetzer, I., Bala, G., von Bloh, W., Feulner, G., Fiedler, S., Gerten, D., Gleeson, T., Hofmann, M., Huiskamp, W., Kummu, M., Mohan, C., Nogués-Bravo, D.,... & Rockström, J. (2023). Earth beyond six of nine planetary boundaries. *Science Advances*, *9*(37). https://www.science.org/doi/10.1126/sciadv.adh2458. Accessed February 20, 2026.

Rockström, J., Steffen, W., Noone, K., Persson, Å., Stuart Chapin III, F., Lambin, E. F., Lenton, T. M., Scheffer, M., Folke, C., Schellnhuber, H. J., Nykvist, B., de Wit, C. A., Hughes, T., van der Leeuw, S., Rodhe, H., Sörlin, S., Snyder, P. K., Costanza, R., Svedin, U., ... & Foley, J. A. (2009). A safe operating space for humanity. *Nature, 461*, 472–475. https://doi.org/10.1038/461472a

Rosenzweig, C., Solecki, W., Romero-Lankao, P., Mehrotra, S., Dhakal, S., & Ali Ibrahim, S. (Eds.). (2018). *Climate change and cities: Second assessment report of the Urban Climate Change Research Network*. Cambridge University Press.

Rosenzweig, C., Solecki, W., Friedman, E., Ramos, G. C. D., Xing, X., Lehmann, M., Tollin, N., Kagan, S. S., Benhallam, J., Dombrov, M., Bader, D., & Pelaccio, N. (2026). Building and using the evidence base for urban climate action: The UCCRN city solutions case study Atlas. *Npj Urban Sustainability, 6*(1), 34. https://doi.org/10.1038/s42949-026-00342-z

Sakschewski, B., Caesar, L., Andersen, L., Bechthold, M., Bergfeld, L., Beusen, A., Billing, M., Bodirsky, B. L., Botsyun, S., Dennis, D., Donges, J. F., Dou, X., Eriksson, A., Fetzer, I., Gerten, D., Häyhä, T., Hebden, S., Heckmann, T., Heilemann, A., ... & Rockström, J. (2025). *Planetary health check 2025: A scientific assessment of the state of the planet*. Potsdam Institute for Climate Impact Research (PIK), 144 p. https://doi.org/10.48485/pik.2025.017

Salvia, M., Pietrapertosa, F., D'Alonzo, V., Clerici Maestosi, P., Simoes, S. G., & Reckien, D. (2023). Key dimensions of cities' engagement in the transition to climate neutrality. *Journal of Environmental Management, 344*, Article 118519. https://doi.org/10.1016/j.jenvman.2023.118519

Schrage, J., & Shrestha, S. (2025). What functions for city networks in local climate governance? *Current Opinion in Environmental Sustainability, 75*, Article 101540. https://doi.org/10.1016/j.cosust.2025.101540

Sharp, E. B., Daley, D. M., & Lynch, M. S. (2011). Understanding local adoption and implementation of climate change mitigation policy. *Urban Affairs Review, 47*, 433–457.

Solecki, W., Delgado, G., Roberts, D., Walsh, B., & Rosenzweig, C. (2021). Accelerating climate research and action in cities through advanced science–policy–practice partnerships. *NPJ Urban Sustainability, 1*, 3. https://doi.org/10.1038/s42949-021-00015-z

Trencher, G., Bai, X., Evans, J., McCormick, K., & Yarime, M. (2014). University partnerships for co-designing and co-producing urban sustainability. *Global Environmental Change, 28*, 153–165. https://doi.org/10.1016/j.gloenvcha.2014.06.009

Ulpiani, G., Vetters, N., Shtjefni, D., Kozarev, V., Dunlop, T., & Pereira, A. G. (2025). Do ambitious cities value collaboration in climate action? Insights from the first group of cities pursuing climate neutrality. *Cities, 162*, 105945. https://doi.org/10.1016/j.cities.2025.105945

Wei, H., & Wei, X. (2024). The role of city network in improving corporate ESG scores: Evidence from the C40 network. *Cities, 152*, 105237. https://doi.org/10.1016/j.cities.2024.105237

PART II

City Innovation Knowledge Gaps and Action Priorities

Chapter 6: Delivery Approaches to Enable Cities to Take Action

Chapter 7: Pillars of Justice and Equity: Advancing Equitable and Just Climate Innovation

Chapter 8: Digitalization for City-Level Models, Data, and Knowledge: Enabling Trust, Transparency, and Evidence-Based Climate Action

The transition from climate ambition to tangible urban transformation requires more than political commitment: it requires robust implementation pathways, equitable processes, and trustworthy information systems. This section explores how the GRAA moves from framework to practice, addressing the fundamental challenge of translating global climate imperatives into locally grounded, inclusive action that serves all inhabitants—human and non-human—across generations, and of broadening local initiatives into global action.

Effective implementation begins with empowering cities as sites of transformative action through multi-level governancestructures, co-production of knowledge across science, policy, and practice, and resilient long-term partnerships. With nods to the original enabling factors in the 2018 GRAA, the authors in this section draw from key issues and themes emerging at the most recent iteration—the 2024 Innovate4Cities conference—to demonstrate through case studies how enabling, equitable, justice-focused pathways strengthen the civic infrastructure and knowledge systems cities need to address the global climate agenda while supporting local decision-making and data-informed frameworks for climate-impact assessment.

Underpinning both pathways and justice frameworks is the infrastructure of data. Digitalization can shift urban efforts from compliance reporting to opportunity-focused, equity-centered action. But this only occurs when such efforts are grounded in data diversity, technology-neutral governance, and local capacity development that transforms fragmented information into unified trustworthy foundations for accelerated climate response.

CHAPTER 6

Delivery Approaches to Enable Cities to Take Action

Tanya Chung-Tiam-Fook, Jayne Engle, Samia Khan, Marnie McGregor, Giorgia Rambelli, Ika Trijsburg, and Maryke van Staden

6.1 Introduction

This chapter presents the three key pathways to implement the Global Research and Action Agenda (GRAA): (a) empowering cities as the foundation for bold, locally grounded climate action; (b) advancing co-production of knowledge across science, policy, and practice; and (c) fostering long-term, cross-sector partnerships. Through these pathways,

T. Chung-Tiam-Fook
7GenCities, Toronto, ON, Canada
e-mail: tanya@7GenCities.org

J. Engle
School of Urban Planning & Max Bell School of Public Policy, McGill University, Montréal, Canada
e-mail: jayne.engle@mcgill.ca

C. Oke et al. (eds.), *A Global Research and Action Agenda on Cities and Climate Change Science*,
https://doi.org/10.1007/978-981-95-6517-7_6

the chapter explores how cities are implementing the GRAA's priorities within municipal governance, innovation systems, and collaborative networks.

The three delivery pathways of the GRAA have evolved over its successive iterations from 2018 to 2024 to move from conceptual framing to implementation systems. Initially, cities were recognized as sites of climate action; today, they are positioned as leaders of systemic transformation (Prieur-Richard et al., 2019). The first pathway, empowering cities for ambitious climate action, has progressed from identifying urban research needs to strengthening institutional readiness, multi-level governance, and access to finance. The second pathway, co-production of knowledge, has shifted from establishing shared research frameworks to embedding co-production within decision-making systems, as a requisite to empower cities and scale action.

S. Khan (✉) · M. McGregor · G. Rambelli
Global Covenant of Mayors for Climate and Energy (GCoM), Brussels, Belgium
e-mail: skhan@gcomprojectsupport.org

M. McGregor
e-mail: mmcgregor@globalcovenantofmayors.org

G. Rambelli
e-mail: grambelli@globalcovenantofmayors.org

G. Rambelli
Urban Transitions Mission (UTM), Global Covenant of Mayors for Climate and Energy (GCoM), Brussels, Belgium

I. Trijsburg
Crawford School of Public Policy, Australian National University, Canberra, Australia
e-mail: itrijsburg@mav.asn.au

M. van Staden
carbonn Climate Center, ICLEI Local Governments for Sustainability, Bonn, Germany
e-mail: maryke.van.staden@iclei.org

The third pathway, partnerships for long-term collaboration, has progressed from a focus on short-term research alliances to the establishment of enduring delivery systems, including city-to-city and regional networks, that bring together cities, national governments, civil society, financiers, and knowledge partners. These systems are increasingly aligned with global mechanisms such as the Intergovernmental Panel on Climate Change (IPCC), Coalition for High Ambition Multi-level Partnerships for Climate Action (CHAMP), Sustainable Urban Resilience for the Next Generation (SURGe), and the Global Stocktake under the Paris Agreement (GST). In doing so, they now embed the GRAA's evolution into an implementation engine for transformational climate action across local, subnational, national, and global scales.

The 2024 GRAA identifies emerging frontiers for research and practice in how cities are designed, governed, and co-created. There is a growing need to embed ecological placemaking, participatory design, and multilevel governance mechanisms that enhance climate resilience while delivering co-benefits for marginalized and vulnerable communities (Irvin et al., 2024). Likewise, incorporating advocates for biodiversity and Indigenous and local land stewards within governance structures through community engagement and participatory planning forums ensures that ecological systems and more-than-human[1] life are meaningfully represented in urban decision-making. Elevating the agency of more-than-human species in city-level dialogs represents a critical step toward truly inclusive and regenerative urban futures.

This chapter explores these needs through examples of research and practice, including Indigenous–municipal–civil society partnerships that illustrate how ecological and participatory approaches are transforming cities into places capable of sustaining all forms of life. The discussion presents these aspects through case examples and concludes by reflecting on the evolving role of cities in shaping climate-resilient and regenerative urban futures designed to sustain the well-being of all beings, human and more-than-human, across generations.

[1] "More-than-human," gestures to the realm of life that not only holds human beings and cultures, but also the exuberant diversity of all natural entities and living systems, with their self-sovereignty and agency. Eco-philosopher David Abram expounds that "the human world is embedded in, permeated by, and dependent upon the more-than-human world that exceeds it" (2024, p. 341). A more-than-human world centers interconnectedness and innate value across all of life.

6.2 Delivery Approaches of the GRAA

This section explores the three implementation pathways of the GRAA.

6.2.1 Empowering Cities for Ambitious Climate Action

Cities can contribute up to 58% of the emission reductions needed to reach net-zero by 2050 if their contributions are embedded in supportive national frameworks (Coalition for Urban Transitions, 2021). Despite this transformative potential, cities face institutional, financial, and data-related constraints that limit their ability to act at scale. Fragmented governance, insufficient fiscal autonomy, and weak integration between local and national climate policies continue to slow progress (IPCC, 2022). The delivery approach of empowering cities for ambitious climate action is explored here through the lens of multi-level governance and coordinated action across city stakeholders.

Multi-level governance—coordinated policymaking across local, subnational, national, and supranational levels—is a foundational enabler for net-zero resilience, and for people- and ecosystem-centered futures (CHAMP, 2024). Effective collaboration across levels yields tangible benefits such as joint planning, shared climate-data infrastructures, and co-financing mechanisms that enable governments to develop climate solutions that are both more equitable and politically and socially durable. Vertical policy integration, in turn, strengthens implementation capacity, enhances buy-in from key stakeholders, and reduces policy fragmentation (IPCC, 2022). Countries with stronger multi-level coordination exhibit more-consistent emission-reduction trajectories and better resource optimization (OECD, 2020).

Local and subnational actors drive progress on the GRAA by acting as the critical bridges between local priorities and national targets. They function as innovation laboratories, piloting mitigation, and adaptation strategies that can replicate and scale national policies and innovation actions. Initiatives such as Mission Innovation's Urban Transitions Mission (Mission Innovation, 2025) illustrate how multi-level governance can support more than 300 cities by 2030 in pioneering people- and ecosystem-centered pathways toward net-zero, resilient futures.

Recent examples demonstrate how effective multi-level governance frameworks are increasingly empowering cities to deliver on national commitments. The latest Nationally Determined Contribution 3.0

submissions reflect a growing formalization of vertical-coordination mechanisms: Brazil has institutionalized "climate federalism" through the Coalition for High Ambition Multi-level Partnerships (CHAMP) initiative; Kenya and Rwanda have integrated subnational actions into NDCs via the Financing Locally Led Climate Action (FLLoCA) program and district-level planning processes; and France ensures vertical consistency through its National Low-Carbon Strategy (SNBC) (CHAMP, 2024; UN-Habitat, 2025).

Strengthening multi-level governance is not merely an ancillary concern; it is a fundamental enabler to empower cities and to deliver climate solutions that are just, scalable, and systemically transformative (IPCC, 2022). Embedding coordination mechanisms, participatory processes, and shared accountability across all levels of governance is essential to implementing science-based targets and advancing the objectives of the Global Stocktake. Local ability to deliver transformative mitigation and adaptation outcomes remains constrained by limited fiscal autonomy, fragmented responsibilities, and restricted access to national and international climate finance—less than 10% of global climate finance currently reaches the local level (CCFLA, 2021). Some cities are overcoming these barriers through innovative financing and partnerships. For example, in Bogotá, the C40 Cities Finance Facility supported the design of 25 km of safe, low-carbon cycle lanes to promote active mobility and reduce transport-related emissions while strengthening climate resilience (C40 Cities Finance Facility, 2022). Similarly, the city of Machala, Ecuador, applied GCF support to implement its "Mangroves for Climate" project, integrating nature-based solutions into coastal infrastructure and community resilience (GCF, 2024). Strengthening fiscal decentralization, aligning budgetary systems with climate priorities, and expanding direct access to finance are essential to empower cities to translate global commitments into tangible, locally driven climate action.

6.2.2 Co-Production of Knowledge

Co-production of knowledge in cities refers to inclusive co-design processes that bring together residents, community groups, municipal actors, and civil-society stakeholders to jointly frame problems, generate solutions, and test interventions. As a complement to multi-level governance and partnership approaches, co-production serves as the connective tissue that links scientific evidence with local experience and institutional decision-making, ensuring that climate actions are contextually grounded, equitable, and collectively owned (Turnhout et al., 2020). Done well, it brings diverse expertise and lived experience to the surface, improves design, and leads to creative, implementable responses with higher legitimacy and uptake.

Establishing early community buy-in and genuine power-sharing, such as through Fung's "democracy cube" framework of participation, authority, and communication, is a prerequisite for durable implementation and trust (Fung, 2006). Co-production aligns with "post-normal" science, where high stakes and contested values require extended peer communities and deliberation (Strand, 2017). When clear, context-based, pluralistic, goal-oriented, and interactive principles guide these participatory processes, the prospects for high-quality co-production of sustainability knowledge are increased (Norström et al., 2020). Importantly, for the co-creation processes and outcomes to be part of long-term pathways of commitment to climate objectives, participatory-futures approaches have provided promising results in multiple community contexts globally (Peach & Smith, 2022).

Knowledge co-production with Indigenous urban communities is becoming an essential practice in global climate research and policy, closely linked to efforts that empower locally grounded action and foster long-term, cross-sector partnerships. Indigenous and other land-based societies, having lived for millennia on the front lines of climatic and ecological change, bring deep place-based expertise that enriches climate research and action across scales. Their insights, grounded in generations of observation, complement technoscientific data and place it in the context of ground truth, offering multi-temporal perspectives often missing in conventional approaches. This knowledge is particularly valuable for mapping intergenerational and future climate scenarios, ensuring that adaptation and resilience strategies are rooted in lived experience and ecological continuity.

Engaging Indigenous, ancestral, and local knowledge systems requires a commitment to collective sensemaking, wisdom, and the integration of diverse forms of expertise. It also demands awareness of historical suppression, particularly in relation to technoscientific paradigms, and of how knowledge systems interact as relational and epistemological frameworks. Teachings from Indigenous contexts offer valuable tools for balancing coexistence between distinct, yet valid, systems, and for cultivating the ethical space where new, decolonizing understandings emerge (Bartlett et al., 2012; Ermine, 2007; Goodchild, 2021). The Mi'kmaq concept of etuaptmumk, or two-eyed seeing, illustrates the power of bridging Indigenous and Western knowledge to guide adaptive, equitable, and life-sustaining climate decisions (Martin et al., 2022). Embracing such relational approaches not only deepens the legitimacy and inclusivity of climate research and governance but also expands the collective capacity to co-produce, imagine, and implement regenerative futures grounded in reciprocity, respect, and shared responsibility.

6.2.3 *Partnerships for Long-Term Collaboration*

Addressing climate change through urban research and innovation demands unprecedented collaboration and coordination across governments and sectors, at speed and scale. The GRAA provides a framework for advancing transition toward low-emission and climate-resilient futures by promoting governance models and institutional arrangements that integrate multiple actors and decision-making levels. Designed to enable coherent responses to interconnected urban and climate challenges, the GRAA represents an emerging model for multi-sectoral collaboration that empowers cities to take evidence-informed, transformative climate action.

Achieving this vision requires long-term collaboration among research institutions, policymakers, city networks, and communities. Such sustained partnerships strengthen collective capacity to align scientific inquiry with policy priorities and practical implementation, ensuring that knowledge remains responsive to evolving urban challenges. Establishing resilient science–policy interfaces is therefore critical to enabling cities to achieve net-zero trajectories while building adaptive capacity over time. The GRAA highlights the strategic role of research and innovation in shaping evidence-based, inclusive, and scalable urban climate solutions. These interfaces must foster co-production of knowledge, continuous feedback loops, and institutional mechanisms that embed both scientific and community-led expertise into decision-making processes (Wyborn et al., 2019).

Research and innovation policies are pivotal to driving systemic transformations through multi-level governance and partnerships. Findings from the Urban Transitions Mission (Mission Innovation, 2025), involving an existing cohort of 117 cities across 46 countries, highlight the urgency of advancing renewable energy, resilient infrastructure, AI-driven solutions, sustainable mobility, efficient buildings, people-centered design, and blended finance models. These priorities demand integrated, transdisciplinary approaches linking technological, institutional, and social innovation. To implement them, cities need robust digital and governance infrastructures, interoperable data systems, and analytical tools for scenario planning. For instance, Amsterdam's Digital Twin integrates real-time data across urban systems to simulate decarbonization pathways, exemplifying how research and innovation can be embedded into governance to enable climate-resilient, future-ready transitions (Wolters, 2023).

Delivering on the GRAA priorities depends on sustained collaboration among governments, researchers, and communities that extends beyond individual projects or funding cycles. Long-term partnerships create continuity across political, institutional, and technical domains, ensuring that innovation systems remain adaptive and responsive to evolving urban challenges. These partnerships serve as critical delivery pathways for implementing the GRAA's vision bridging research, policy, and practice to scale climate solutions and maintain momentum over time (Hadfield et al., 2021).

In light of this, innovation must not only accelerate technological deployment but also ensure context-appropriate solutions accessible to cities of all sizes and income levels. Urban transitions must be grounded in science while incorporating the lived experience, feedback, and expertise of city practitioners. Researchers, universities, and innovators must collaborate with local and subnational governments to co-create fit-for-purpose solutions that place people at the center of bold, long-term transformations that can ensure that climate action is inclusive, effective, and durable.

Data-informed decision-making is a cornerstone of effective climate governance and an essential principle for advancing inclusive, evidence-led urban transitions. Yet, as cities embed evidence-driven approaches into local climate strategies, they operate within an increasingly complex landscape of declining trust and increasing spread of misinformation. The rise of climate disinformation—the deliberate dissemination of false or misleading narratives aimed at denying, delaying, or distorting climate action—poses a significant challenge to evidence-based policymaking and public confidence (IPIE, 2025). The World Economic Forum's Global

Risks Reports in 2024 and 2025 identified disinformation and misinformation as the top global risks of the immediate term, reflecting both their scale and their potential to undermine collective action (WEF, 2024, 2025).

Climate disinformation seeks to deny and delay climate action, using techniques from falsifying scientific findings to advancing conspiratorial explanations for climate events and disasters, targeting both policymakers and the public (Perl et al., 2018). Usually driven by political or economic interests, disinformation has demonstrable impact on cities' ability to pursue decisive, evidence-based climate action in three key ways: first, it pollutes the information environment with inaccuracies, making evidence-based policy increasingly challenging; second, it shapes public sentiment through false narratives that amplify polarization and erode consensus on climate action; and third, it undermines institutional trust and collective will, diminishing social license and public willingness to support or participate in local climate initiatives. (UNESCO, 2022).

In the context of data-driven decision-making, cities increasingly recognize that combating disinformation is not just a communication issue but a governance and resilience challenge. New forms of targeted support are emerging to help local governments address this threat, including initiatives by C40 Cities and ICLEI, and the development of the Disinformation in the City Response Playbook (C40 Cities, 2025; ICLEI Canada, n.d.; Trijsburg et al., 2024). Co-created by researchers, municipal practitioners, and experts from civil society, technology, and media, the Playbook offers an adaptable framework for understanding community fault lines, preventing and debunking misinformation, and investing in community recovery to build collective resilience to future disinformation.

Embedding such frameworks within municipal innovation ecosystems reinforces cities' capacity to harness data effectively, anticipate risk, and maintain public confidence in science-based decision-making. These approaches exemplify how knowledge co-production and long-term partnerships, which are GRAA core delivery pathways, extend beyond technological innovation to include the social and informational infrastructures that sustain inclusive climate governance.

6.3 Case Examples of "Edge Innovation" Delivery Approaches

The three cases discussed in this chapter are intentionally selected as edge innovations with multi-solving outcomes. The *CityStudio* case provides a systemic infrastructure for ongoing municipal-community-university co-creation and delivery of climate-resilience objectives, and has the benefit of extensive youth engagement, drawing together transdisciplinary teams of students from across regional universities. The second example, the *Open Space Tool for Embedding Resilience and Reconciliation,* provides an assessment and measurement framework for urban green space, and the *FreeLand* and *FreeHouse Proofs of Possibility* lay the grounding for climate-related systems change, through alternatives to land and housing ownership. Cases two and three are informed by Seven Generation Cities, an initiative co-led with Indigenous communities (see Engle et al., 2022, xvii; 7gencities, 2024).

Case Example 1: Multi-Level Collaboration Through CityStudio and The Kali Polis Project An illustrative example of long-term collaboration translating research and innovation into practice is CityStudio Vancouver, an innovative civic engagement and learning lab that connects municipal government, universities, and community partners to co-create local climate and sustainability solutions. Founded by the City of Vancouver in partnership with Simon Fraser University, the University of British Columbia, Langara College, and the British Columbia Institute of Technology, CityStudio provides students with opportunities to co-design and implement real-world urban projects in areas such as green infrastructure, the energy transition, and climate resilience, earning academic credit while contributing to municipal priorities.

Working under the strategic direction of the Mayor and Council and aligning with local policy frameworks, interdisciplinary teams of city staff, students, and faculty collaboratively develop experimental projects that address Vancouver's pressing sustainability and inclusion challenges. Each year, these projects are presented at a public showcase at City Hall, fostering transparency, civic engagement, and knowledge-sharing between residents, elected officials, and city staff. Since its inception, over 9500 students, 450 city staff, and nearly 500 university faculty have collaborated on more than 2400 local projects, inspiring a growing global network of 14 CityStudios across Canada and beyond, collectively engaging more than 14,000 students and 700 city staff in 2600 projects.

Building on this model, The Kali Polis Project, launched at the Demos Center of the American College of Greece in Athens, with support from the Global Covenant of Mayors for Climate and Energy (GCoM), adapts the CityStudio approach to a Mediterranean and European context. The program serves as an urban studio and internship platform that connects Greek undergraduate and graduate students with municipal authorities to co-design and implement projects addressing climate priorities such as heat resilience, wildfire management, and flood mitigation. Starting in 2026, the initiative aims to strengthen city-campus collaboration as a form of embedded governance innovation, where universities act as living laboratories for sustainable transformation and youth engagement.

Together, CityStudio Vancouver and The Kali Polis Project exemplify how long-term, cross-sectoral partnerships between academia, local governments, and civil society can implement the GRAA's delivery pathways. They demonstrate how co-production of knowledge, grounded in local realities yet informed by global collaboration, can generate scalable solutions that bridge research, policy, and practice, ensuring that cities remain at the forefront of climate innovation.

Situating Case Examples 2 and 3: Seven-Generation Thinking and Effective Social Infrastructures for Climate Action

The political, economic, climatic, and social challenges confronting communities, bioregions, and cities are profound and interconnected. A growing body of global scientific evidence complemented by the lived experience and local expertise of affected communities demonstrates that climate vulnerability and environmental degradation are closely linked to structural inequalities and unsustainable patterns of development embedded within prevailing institutional and economic systems. Addressing these challenges requires rethinking current models of growth and governance to ensure that future transitions are equitable, regenerative, and grounded in justice (Engle et al., 2022).

Shifting systems and mindsets, lifting collective ambition, and seeding resilient climate futures for all peoples and bioregions is a massive undertaking. However, there exist diverse ancestral and modern wisdoms and approaches to innovation that can inform, and be centered in and inspired by, seven-generations thinking and practice, and the inclusive futures that are built from here. Innate to many Indigenous and other ancestral cultures and Local Knowledge are sacred agreements concerning reciprocal land relationships and regeneration, systems and relational complexity thinking, and governance and regenerative economies. These ancestral approaches are not about returning to a static and traditional past; they are a set of relational and collectively held teachings, commitments, and activations of how to live well and with generosity and dignity, with each other and with the Earth (Big Canoe et al., 2025). They provide a living framework for cities to organize, provision, and govern social systems and communities according to a nested set of coherent circles of interconnected relationships and flows of energy and intelligence (human and more-than-human).

Weaving the deep learning and practice of ancestral systems backed by centuries, if not millennia, of evidence-based practice with emerging climate and community resilience research and local actions is an effective way of applying two-eyed seeing in resilience and transformational adaptation strategies to address increasingly complex climate threats (Chung-Tiam-Fook, 2015). These systems have long been present, showing up through land relationships and stewardship, urban food forests, land and housing commons, climate justice, land and bioregional governance, community finance and circular economies, care infrastructure, peer-to-peer platforms, and pro-social technologies; they create a framework for building resilient, seven-generations cities by acting on foundational spiritual and social values (Chung-Tiam-Fook et al., 2022).

The 7GenCities collaborative of Indigenous, local, civic, and municipal partners draws on diverse knowledges and approaches, including foundational wisdom inspired by the Haudenosaunee Seventh Generation Principle (Haudenosaunee Confederacy, n.d.) and similar Indigenous teachings, to reimagine and co-build seven-generation futures for communities, cities, and bioregions based on transitioning toward life-affirming and Reconciliation-embedded logics and infrastructures. The Seventh Generation Principle situates current humans as part of an intergenerational arc where people are called to learn from the quantum wisdom of past ancestors and the Earth as they plan, make decisions, and provision for current and future generations; this includes climate-action planning (Hill et al., 2021, cited in Mortillaro, 2021).

Recognizing the agency and value of all peoples, natural systems, and future generations enables seven-generation cities to model how to be better stewards of land and food systems, build resilient civic infrastructures and economies, and govern in ways that sustain life across generations.

Intergenerational collaboration is vital to building regenerative, resilient, and inclusive cities. It demands co-leadership across sectors and the exchange of wisdom across generations, cultures, and lands. By weaving together scientific, Indigenous, and local, community-based knowledge, such collaboration strengthens the legitimacy and inclusiveness of climate governance. Multigenerational decision-making ensures diverse priorities guide action, recognizing that younger generations inherit the impacts of unsustainable development. Today's leaders bear a responsibility to enable just and regenerative transitions, while youth stewards hold institutions accountable to their commitments, and their inaction.

The following cases take seven-generation thinking into account and aim to demonstrate GRAA key approaches. The book *Embedding Resilience and Reconciliation Tool for Open Spaces* (Chung-Tiam-Fook & Valentini, 2025) fosters long-term relationships and participatory planning among a diversity of open-space actors to collect data and improve climate and community resilience in open spaces. The tool supports co-production of local, Indigenous, and technoscientific knowledges to inform inclusive policies, and empowers open-space stewards with multifaceted data and climate and community strategies. The emerging FreeLand and FreeHouse initiatives promote knowledge co-production and transsystemic Indigenous-federal-global legal frameworks in the Canadian context while fostering long-term partnerships among municipalities, Indigenous urban communities, scholars, and practitioners. These multisolving infrastructures demonstrate how land and housing development can be regenerative and not extractive, climate-resilient, community-stewarded, and beneficial to the commons.

Case Example 2: Participatory Engagement Methods: The Embedding Resilience and Reconciliation Tool for Open Spaces Chung-Tiam-Fook and Valentini's (2025) *Embedding Resilience and Reconciliation: Tool for Open Spaces* (enables municipal, civic, Indigenous, and other local public-space stewards and placemakers with the data-collection and decision-making capabilities to apply targeted nature-based solutions and community outcomes to improving open public spaces. The tool provides a place-based strategy for monitoring, evaluating, and enhancing open spaces about climate-resilience capacity, including improving the capacity to withstand and recover from impacts such as flooding and wildfires. The central aim is to improve community well-being by fostering inclusive, sustainable, resilient, and safe environments and participatory planning and governance approaches that enable the collective flourishing of both communities and more-than-human worlds.

The tool integrates mixed-methods data collection and evaluation frameworks grounded in participatory and Indigenous approaches to monitoring and governance. It engages diverse open-space stewards, including Indigenous and local communities, whose place-based expertise in areas such as watershed and flood management helps reveal ground truths for scientific data. Data and knowledge shared by Indigenous and cultural Knowledge Keepers are managed according to Indigenous data sovereignty and governance principles (e.g., OCAP Principles for data sovereignty; CARE Principles for data governance; FAIR Principles for open data) and collected, stored, and analyzed securely under Indigenous guidance, with access limited to designated project leads.

A unique qualitative framework and metric symbolizing a canoe journey guides participants in processes of righting relationships with Indigenous Peoples and the land, transforming public spaces, and fostering inclusive, place-based stewardship by diverse communities. The tool acknowledges that each allied steward's journey is a reflective, iterative, and dynamic progression through key stages of: (1) improved recognition of the value of engaging Indigenous perspectives; (2) emerging awareness; (3) exploration and learning; (4) community engagement and integration of Indigenous inclusive approaches; and 5) co-creation with Indigenous Community leadership in shaping climate and community resilient public spaces. Truth and Reconciliation questions are embedded into climate resilience and community-well-being criteria to hold cities accountable for decolonizing public spaces, and centering Indigenous Knowledges, more-than-human perspectives, and future generations in resilience planning.

The public-space tool was piloted at Evergreen Brick Works, a multi-use public space in Tkaronto (Toronto) with natural trails, ponds, urban forest, gardens, green playgrounds, community markets, and gathering and convening spaces. The Brick Works—the site of a former brick foundry that is based on First Nations treaty and ancestral lands—is one of Canada's most important post-industrial adaptive-reuse projects. Piloting the tool on site with diverse First Nations leaders and practitioners enriched its value, making it foundational to Indigenous engagement and Reconciliation-centered climate and community resilience action. Feedback from Indigenous advisors highlighted the tool's cultural and strategic relevance, supporting Indigenous land co-stewardship, urban land sovereignty, and community climate resilience.

Case Example 3: Systemic Approaches to Addressing Property Challenges: FreeLand and FreeHouse Models of Land and Housing Stewardship Matters of property and ownership fundamentally shape land and housing tenure, urban economies, and equitable access. As housing crises deepen and local economies face growing precarity, it is imperative to challenge the systems and assumptions that sustain inequality. Emerging initiatives such as FreeLand and FreeHouse reimagine dominant property regimes by drawing on Indigenous Knowledge systems and seven-generation principles. They introduce alternative, regenerative models of land and housing governance that move beyond extractive paradigms toward community stewardship, reciprocity, and long-term resilience offering a glimpse into how cities might redefine ownership and value in pursuit of equitable and sustainable urban futures.

Beyond the legal and philosophical assertions of the Rights of Nature movement, FreeLand proposes to create a new legal mechanism and governance framework aligned with Indigenous and other local legal traditions and place-based forms of lawing and stewardship, for land to be legally recognized as a self-sovereign agent. In terms of knowledge co-production, FreeLand creates space and a transsystemic approach for Indigenous and other forms of earth jurisprudence legal scholarship that challenge the normative ethos of land as property and generate new possibilities for legal pluralism in Canada. This "legal braiding" will be increasingly relevant as efforts to decolonize and reshape cities and civic institutions that had their genesis in the days of colonial rule increase in number and scope.

The FreeLand project currently under development explores self-owning land: land as common-good infrastructure that is held in non-ownership trust, and where the land's legal, social, and economic governance operates under an Indigenous legal order, enabled by a transsystemic strategy. It is a space where land is not "owned" in a conventional sense but related to and respected as a living entity with agency in accordance with various Indigenous worldviews (depending on the land context and First Nations stewards).

Self-ownership of land provides a strategic container for understanding self-owning housing, or FreeHouse, enabling buildings to exist on land outside of speculative real-estate markets, yet within resilient systems of care and stewardship.[2] FreeHouse offers a framework for non-ownership-based tenure, grounded in resident governance, relational accountability, and Indigenous-led land protocols. It invites a shift from enclosure to reciprocity and from ownership to stewardship. It explores how the value flows of developing land can be regenerative and not extractive, and how its value can be stewarded for the sustainable benefit of the commons.

FreeHouse Tkaronto, currently in its participatory design stage, advances Indigenous self-determined housing and land governance through long-term partnerships among Indigenous Knowledge Keepers, municipal leaders, and civic practitioners. The initiative envisions housing as a shared commons, where Indigenous stewards, residents, and public authorities co-create mutual value through commons-based governance, resident–stewardship agreements, and holistic outcome frameworks reflecting shared care and benefit. Located on degraded land along Toronto's ravine ecosystems, FreeHouse proposes regenerative, affordable housing where residents engage in land stewardship inspired by Indigenous practices.

As a proof of possibility, it integrates climate resilience, circular resource management, social equity, and ecological repair while advancing Indigenous-led legal pluralism and community-led governance. Together, these elements model a new relational approach to housing that redefines ownership, restores ecosystems, and strengthens the interdependence between people and land. By designing a housing model that intentionally addresses ecological repair, social equity, and legal resurgence, FreeHouse becomes a multi-solving demonstration: an infrastructure designed not only for shelter, but for climate resilience and justice.

6.4 Conclusion

The evolution of the GRAA marks a shift from framing research priorities to putting systems into practice that enable transformative climate action across scales. Through its three delivery pathways—empowering cities, co-producing knowledge, and fostering long-term partnerships—the GRAA

[2] FreeLand and FreeHouse initiatives have been developed with Dark Matter Labs in partnership with Indigenous communities and multiple partners (more information is available in Dark Matter Labs (2025) and at darkmatterlabs.property).

functions as an implementation framework connecting science, policy, and practice.

The case examples in this chapter exemplify how GRAA pathways converge in practice. CityStudio Vancouver and the Kali Polis Project show how long-term, cross-sector partnerships between municipalities, universities, and communities form institutions of co-production and strengthen civic-innovation ecosystems. The Embedding Resilience and Reconciliation Tool for Open Spaces demonstrates how approaches that are participatory and co-led by Indigenous and other local participants can embed reconciliation and ecological stewardship within municipal decision-making, making climate resilience both measurable and relational. The FreeLand and FreeHouse initiatives extend this further, reimagining land and housing as shared commons and showing how Indigenous legal frameworks, resident stewardship, and ecological regeneration can transform extractive systems into regenerative ones.

The cases demonstrate that empowering cities requires more than technical capacity: it demands relational governance, shared accountability, and a recognition of urban resources and communities as co-actors in resilience. By embedding multi-level coordination, inclusive knowledge systems, and enduring partnerships within local innovation ecosystems, cities can deliver climate solutions that are just, regenerative, and scalable.

The GRAA's delivery pathways demonstrate that transformational climate action emerges where science meets stewardship, policy meets participation, and innovation is guided by justice and care, laying the foundation for seven-generation cities capable of sustaining human and more-than-human life well into the future. Through the practical cases explored, this chapter demonstrates that realizing such transformation requires more than intent; it demands reflection on how the GRAA itself is implemented across sectors and contexts. It calls for critical inquiry, including: where do enabling powers, data, and finance reside within cities, and where do they not? Who defines problems and success metrics? How are Indigenous and Local Knowledges centered, safeguarded, and valued? And which alliances are formalized, resourced, and capable of learning across cycles of change? Only by addressing these questions can the GRAA continue to evolve as a living framework, one that not only accelerates climate action, but also nurtures the ethical, relational, and regenerative foundations of a just, climate-resilient urban future.

Competing Interests The authors have no conflicts of interest to declare that are relevant to the content of this chapter.

References

7GenCities. (2024, March). *7GenCities learning report*. https://cdn.prod.website-files.com/6462104f72fa898a55556e56/660da9644f94a4e7714e8dfe_Report_7GenCities_Learning-Gathering_compressed.pdf

Abram, D. (2024). On the origin of the phrase "More-Than-Human." In C. Rodríguez-Garavito (Ed.), *More than human rights: An ecology of law, thought, and narrative for earthly flourishing*, NYU Law.

Bartlett, C., Marshall, M., & Marshall, A. (2012). Two-eyed seeing and other lessons learned within a co-learning journey of bringing together indigenous and mainstream knowledges and ways of knowing. *Journal of Environmental Studies and Sciences, 2*, 331–340.

Big Canoe, B., Rayna, A., & Hamilton, D. (2025). *The Kinship Way*. Kindred Spirits. https://kindredspirits.earth/wp-content/uploads/2025/08/Kinship-Way-Manuscript-4.pdf

C40 Cities Climate Leadership Group, Inc. (2025). *Disinformation*. https://www.c40.org/what-we-do/building-a-movement/disinformation/

C40 Cities Finance Facility. (2022). *CFF impact brief | Cycle Avenue Medio Milenio: Colombia's first large-scale cycle avenue* [Impact brief]. C40 Cities Finance Facility. https://cff-prod.s3.amazonaws.com/storage/files/0at55am16hjwEIxhHugVathwmDyXZg7rz03PzPye.pdf

Chung-Tiam-Fook, T., & Valentini, S. (2025). *Embedding resilience and reconciliation: Tool for open spaces*. Evergreen. https://evergreen.ca/resource-hub/?post_type=resource&p=7388

Chung-Tiam-Fook, T., Engle, J., & Agyeman, J. (2022). Awakening seven generation cities. In J. Engle, J. Agyeman, & T. Chung-Tiam-Fook (Eds.). *Sacred cities: Building seven generation cities*. Routledge.

Chung-Tiam-Fook, T. (2015). Transformational processes for community-focused adaptation and social change: A synthesis. *Climate and Development, 9*(1). https://doi.org/10.1080/17565529.2015.1086294

Cities Climate Finance Leadership Alliance (CCFLA). (2021). *The state of cities climate finance 2021*. Cities Climate Finance Leadership Alliance. https://citiesclimatefinance.org/publications/2021-state-of-cities-climate-finance

Coalition for High Ambition Multilevel Partnerships. (2024). *Summary of activities report 2024*. https://www.cities-and-regions.org/wp-content/uploads/champ_summary_of_activities_report_2024.pdf

Coalition for Urban Transitions. (2021). *Seizing the urban opportunity: How national governments can recover from COVID-19, secure shared prosperity and tackle the climate crisis through cities*. Coalition for Urban Transitions. https://urbantransitions.global/en/publication/seizing-the-urban-opportunity/

Dark Matter Labs. (2025). FreeHouse Tkaronto. Blog written by C. Po with T. Chung-Tiam-Fook, J. Engle, and C. Nolan. *Medium*. https://provocations.darkmatterlabs.org/freehouse-tkaronto-4e6c2d27967a

Engle, J., Agyeman, J., & Chung-Tiam-Fook, T. (Eds.) (2022). *Sacred civics: Building seven generation cities*. Routledge.

Engle, J., Britton, T., & Glode-Derochers, P. (2022). Social infrastructure for our times. In J. Engle, J. Agyeman, & T. Chung-Tiam-Fook (Eds.), *Sacred civics: Building seven generation cities*. Routledge.

Ermine, W. (2007). The ethical space of engagement. *Indigenous Law Journal, 6*(1), 193–203.

Fung, A. (2006). Varieties of participation in complex governance. *Public Administration Review, 66*(S1), 66–75. https://doi.org/10.1111/j.1540-6210.2006.00667.x

Goodchild, M. (2021). Relational systems thinking. *Journal of Awareness-based Systems Change, 1*(1). https://jabsc.org/index.php/jabsc/issue/view/76

Green Climate Fund. (2024). *FP235: Mangroves for climate: Public, private and community partnerships for mitigation and adaptation in Ecuador* [Project approval]. https://www.greenclimate.fund/project/fp235

Hadfield, P., Oke, C., & Verbeeck, J. (2021). *Regional research and innovation for city climate action: Global synthesis report*. Connected Cities Lab, University of Melbourne. https://doi.org/10.26188/14743125

Haudosaunee Confederacy. Values: Seventh Generation Principle. https://www.haudenosauneeconfederacy.com/values/

Hill, R. Interview, In Mortillaro, N. (2021, January 21). How climate action can benefit from Indigenous tradition of '7th-generation decision-making.' *What on Earth*, CBC News. https://www.cbc.ca/news/science/what-on-earth-indigenous-seventh-generation-thinking-climate-action-1.5882480

ICLEI Canada. (n.d.). *Responding to climate misinformation: FAQs and resources for local governments*. https://icleicanada.org/project/misinformation/

Intergovernmental Panel on Climate Change. (2022). *Climate change 2022: Impacts, adaptation and vulnerability. Contribution of working group II to the sixth assessment report of the intergovernmental panel on climate change*. Cambridge University Press. https://doi.org/10.1017/9781009325844

International Panel on the Information Environment (IPIE). (2025). *Facts, fakes, and climate science: Recommendations for improving information integrity about climate issues* (SFP2025.2). International Panel on the Information Environment. https://doi.org/10.61452/QHRL3301

Irvin, A., Oke, C., Khan, S., Jance IV, B., Ranalder, L., Barth, B., Deacon, A., Abusamra, R., Addison, A., Adebisi, J., Adsett, C., Ailio, E., Al Jadaa,

D., Aslan, S., Assalini, S., Assarkhaniki, Z., Astolfo, J. P., Badino, M., Bai, X., …, & Williams, A. (2024). *Next-generation city climate action: Findings from the 2024 Innovate4Cities conference and update to the global research and action agenda on cities and climate change science (GRAA)*. Global Covenant of Mayors for Climate and Energy, UN-Habitat & University of Melbourne. https://doi.org/10.26188/27147612

Martin, D. H., Marshall, A., & Marshall, M. (2022). Two-eyed seeing: Embracing the power of Indigenous knowledge for a healthy and sustainable ocean. *PLOS Biology, 20*(10). e3001876. https://doi.org/10.1371/journal.pbio.3001876

Mission Innovation. (2025). *Urban Transitions Mission (UTM)*. https://mission-innovation.net/our-missions/urban-transitions-mission

Norström, A. V., Cvitanovic, C., Löf, M. F., West, S., Wyborn, C., Balvanera, P., Bednarek, A. T., Bennett, E. M., Biggs, R., de Bremond, A., Campbell, B. M., Canadell, J. G., Carpenter, S. R., Folke, C., Fulton, E. A., Gaffney, O., Gelcich, S., Leach, M., Le Tissier, M., & Österblom, H. (2020). Principles for knowledge co-production in sustainability research. *Nature Sustainability, 3*(3), 182–190. https://doi.org/10.1038/s41893-019-0448-2

Organisation for Economic Co-operation and Development (OECD). (2020). *A territorial approach to the sustainable development goals: Synthesis report*. OECD Publishing. https://doi.org/10.1787/e86fa715-en

Peach, K., & Smith, L. (2022). Participatory futures: Reimagining the city together. In J. Engle, J. Agyeman, & T. Chung-Tiam-Fook (Eds.), *Sacred civics: Building seven generation cities*. Routledge.

Perl, A., Howlett, M., & Ramesh, M. (2018). Policy-making and truthiness: Can existing policy models cope with politicized evidence and willful ignorance in a "post-fact" world? *Policy Sciences, 51*(4), 581–600. https://doi.org/10.1007/s11077-018-9334-4

Prieur-Richard, A.-H., Walsh, B., Craig, M., Melamed, M. L., Colbert, M., Pathak, M., & Bai, X. (2019). Global research and action agenda on cities and climate change science. *Urban Climate, 29*, Article 100482. https://doi.org/10.1016/j.uclim.2019.100482

Strand, R. (2017). Post-normal science. In C. L. Spash (Ed.), *Routledge handbook of ecological economics* (pp. 288–297). Routledge.

Trijsburg, I., Sullivan, H., Park, E., Bonotti, M., Costello, P., Nwokora, Z., Pejic, D., Peucker, M., & Ridge, W. (2024). *Disinformation in the city: Response playbook*. University of Melbourne. https://doi.org/10.26188/26866972

Turnhout, E., Metze, T., Wyborn, C., Klenk, N., & Louder, E. (2020). The politics of co-production: Participation, power, and transformation. *Current Opinion in Environmental Sustainability, 42*, 15–21. https://doi.org/10.1016/j.cosust.2019.11.009

UN-Habitat. (2025). *Urban content in NDC 3.0—A global snapshot for COP30. United Nations Human Settlements Programme* [Lead authors: Patrick Geoghegan, Lea Ranalder, Nicola Tollin]. https://unhabitat.org/sites/default/files/2025/11/urban_content_in_ndc_3.0._a_global_snapshot_for_cop30.pdf

United Nations Educational, Scientific and Cultural Organization (UNESCO). (2022). *Journalism, "fake news" and disinformation: A handbook for journalism education and training*. UNESCO. https://unesdoc.unesco.org/ark:/48223/pf0000379386

Wolters, B. M. (2023). *Exploring the potential of urban digital twins in climate adaptive development: A case study research on the Gnephoekpolder, the Netherlands* [Master's thesis, Wageningen University & Research & Delft University of Technology]. AMS Institute. https://resolver.tudelft.nl/1242296b-596e-484f-82c9-9ffb6d949cd8

World Economic Forum. (2024). *Global risks report 2024*. World Economic Forum. https://www.weforum.org/publications/global-risks-report-2024

World Economic Forum. (2025). *Global risks report 2025*. World Economic Forum. https://www.weforum.org/publications/global-risks-report-2025

Wyborn, C., Datta, A., Montana, J., Ryan, M., Leith, P., Chaffin, B., Miller, C., & Van Kerkhoff, L. (2019). Co-producing sustainability: Reordering the governance of science, policy, and practice. *Annual Review of Environment and Resources, 44*(1), 319–346. https://doi.org/10.1146/annurev-environ-101718-033103

CHAPTER 7

Pillars of Justice and Equity: Advancing Equitable and Just Climate Innovation

Juliet Aluoch Oluoch, Julie Greenwalt, Ashali Bhandari, Emily Bolo, Jazmin Burgess, Krisztina Campbell, Asma Jhina, Dave Mwita Jomo, Smruti Jukur, Florence Lozet, Clara Marais, Ben Odongo, Sheela Patel, Paty Romero-Lankao, Caterina Sarfatti, and Meggan Spires

7.1 Introduction

Cities are pivotal arenas where the intersecting challenges of climate risk, vulnerability, low-carbon transitions, and justice are most acutely manifested. Urban populations, particularly low-income, informal, racialized,

J. A. Oluoch
Ufanisi Research Network, Nairobi, Kenya
e-mail: oluochjuliet08@gmail.com

J. Greenwalt (✉) · F. Lozet
Cities Alliance, Brussels, Belgium
e-mail: jgreenwalt@citiesalliance.org

F. Lozet
e-mail: florencel@unops.org

C. Oke et al. (eds.), *A Global Research and Action Agenda on Cities and Climate Change Science*,
https://doi.org/10.1007/978-981-95-6517-7_7

and otherwise marginalized groups, bear the heaviest burden of climate change despite contributing least to global emissions (IPCC, 2022). As urban climate hazards intensify, there is an urgent need to reconfigure

---

A. Bhandari
Mumbai, India

E. Bolo
University of Nairobi, Nairobi, Kenya

J. Burgess · K. Campbell
C40 Cities, London, UK
e-mail: jburgess@c40.org

K. Campbell
e-mail: kcampbell@c40.org

A. Jhina
Global Covenant of Mayors, London, UK
e-mail: ajhina@globalcovenantofmayors.org

D. M. Jomo
Ufanisi Research Network, Urban Green Consultants, Nairobi, Kenya

S. Jukur · S. Patel
Society for the Promotion of Area Resource Centres, Mumbai, India

C. Marais · B. Odongo · M. Spires
ICLEI Africa, Cape Town, South Africa
e-mail: clara.marais@iclei.org

B. Odongo
e-mail: ben.odongo@iclei.org

M. Spires
e-mail: meggan.spires@iclei.org

P. Romero-Lankao
Department of Sociology, University of Toronto Scarborough, Scarborough, Canada
e-mail: paty.romerolankao@utoronto.ca

C. Sarfatti
C40 Cities, Milan, Italy
e-mail: csarfatti@c40.org

innovation processes to avoid reinforcing existing socio-spatial inequalities (Romero-Lankao et al., 2023a, 2023b; Ziervogel et al., 2017). A justice approach to the adoption of innovations and technologies, economic sector transitions, and expansion of infrastructure and service coverage that reduces emissions while equitably meeting the needs of all city residents, especially those historically marginalized, is also crucial.

This chapter is organized according to three interdependent justice dimensions widely used in urban climate scholarship: distributive (fair allocation of risks and benefits), procedural (inclusive and accountable decision-making), and recognition (the legitimate inclusion of diverse identities and epistemologies) (Bulkeley et al., 2014; Schlosberg, 2007). These dimensions map closely onto the GRAA pillars updated through the Innovate4Cities discussions: access and reliability, sufficiency and sustainable consumption, participation and engagement, informality, gender, adaptation and mitigation, conflict and crisis response, Indigenous Knowledges and decoloniality, and youth and intergenerational justice (UN-Habitat, 2024).

The chapter analyzes empirical innovations across a deliberately diverse set of urban contexts (large metropolises and smaller secondary cities in both the Global North and South) to identify enabling conditions and tradeoffs in justice-oriented climate action. Case studies include community energy hubs and microgrids in Kibera (Nairobi), clean-cooking stoves in Kampala, Barcelona's Energy Advice Points, LA100 equity strategies, Mexico City's water-harvest programs, youth-led planning in Mumbai, informal sector inclusion in Accra, resilience programming in informal settlements in Beira, the integration of Indigenous Knowledge in water management in Cuenca, and climate-resilient, inclusive displacement-integration programs in Atbara and Jigjiga. These examples were chosen to illustrate distinct justice challenges and mechanisms, rather than to provide exhaustive coverage.

Across these contexts, three practical themes recur. First, local recognition and formal acknowledgement of informal or community-led systems (from distributed energy to local disaster committees) unlocks both distributive benefits and governance legitimacy. Second, participatory processes that are resourced and institutionally embedded (not tokenistic) enable procedural justice and improve the uptake and durability of interventions. Third, equity-sensitive metrics and financing, including targeted subsidies, catalytic grants, and capacity funding for municipal governments, are essential to avoid regressive outcomes (e.g., green gentrification) and to scale sufficiency-focused measures (O'Neill et al., 2018; Raworth, 2017).

7.2 Distribution: Equity in Access and Resource Allocation

7.2.1 *GRAA Pillars: Access and Reliability; Sufficiency; Sustainable Consumption and Production*

The principle of distributive justice must drive all climate change mitigation and adaptation actions, across planning, implementation, and monitoring, to safeguard communities and fairly allocate the burdens and benefits of climate actions (Reckien et al., 2025). By delivering urban climate action that redresses longstanding inequities in access to basic services, local governments have an opportunity to lift residents out of poverty, provide better livelihoods for all, and increase resilience to intensifying climate impacts (C40, 2019).

The accessibility and affordability of basic services like energy, clean water, sanitation, and affordable housing are key to meeting people's most basic needs. Inequalities in access to services are especially acute for the 1 billion people living in informal settlements (UN-Habitat, 2025). Systemic inequities and vulnerability are further exacerbated by the impacts of the climate crisis on urban infrastructure, food systems, social provisioning systems, and the adaptive capacity of communities (IPCC, 2022).

Sufficiency refers to the equitable provision of essential services and goods that meet basic human needs without excessive consumption or environmental degradation (Weindl et al., 2022); it intersects with sustainable consumption and production. The concept responds directly to challenges identified in the Seventh Assessment Report (AR7) scoping process, which emphasized consumption patterns as a cross-cutting theme in sustainable urban transitions (IPCC, 2025).

From an equity perspective, ensuring sufficiency demands that all urban residents have reliable access to climate-resilient services at levels that maintain health and dignity, while avoiding over-use that exacerbates ecological crises (Raworth, 2017). A just transition recognizes that affluent urban populations often exceed planetary boundaries, intensifying greenhouse gas emissions and resource depletion, while those in marginal communities, who contribute least to emissions, disproportionately bear the impacts.

The 2024 Innovate4Cities Outcomes Report identifies sufficiency-driven measures such as zero-waste urban planning, shared mobility, and circular economy policies as critical pathways for cities to reduce emissions while advancing social equity (UN-Habitat, 2024). Such measures align with AR7's emphasis on transformative-consumption governance, yet current urban governance models seldom integrate sufficiency measures with distributive justice goals, often privileging neoliberal growth paradigms over redistributive equity (IPCC, 2025; Bulkeley et al., 2014).

Cities need to institute sufficiency through regulatory frameworks (e.g., urban-density mandates, per-capita water caps) alongside incentives for low-resource-use behaviors (e.g., safe heat metering, reuse hubs) (Geels et al., 2018) and apply them at scale. Equitable consumption policies also necessitate coupling environmental limits with social safety nets, such as progressive pricing scales and benefit transfers, to avoid imposing burdens on low-income groups (Spash & Conway, 2021).

Creating frameworks of sufficiency and sustainable consumption that use quantitative indicators (e.g., resource footprints, the greenhouse gas intensity of services) with qualitative metrics of well-being, justice, and capabilities (O'Neill et al., 2018) can help decision-makers evaluate trade-offs between efficiency and equity, ensuring that the benefits of climate action are distributed fairly across all urban residents.

Cities worldwide—as demonstrated in the following case studies from **Kampala, Kibera (Nairobi), Barcelona, Los Angeles, and Mexico City**—are leading efforts to address challenges like energy poverty and climate impacts through equitable access to resources and resilient infrastructure. From distributed energy systems in informal settlements to participatory waste and adaptation initiatives, these examples highlight how localized, sufficiency-driven approaches can both reduce greenhouse gas emissions and strengthen service reliability for vulnerable urban populations. Cities need more resources, institutional capacity, and devolved authority to scale such climate-justice innovations at the pace and depth required (C40 Cities, 2022a).

7.2.2 *Case Studies*

Los Angeles: Inclusive Engagement for an Equitable and Just Energy Transition

The LA100 Equity Strategies, sponsored by the Los Angeles Department of Water and Power, drew on community-based organization (CBO) engagement in underserved neighborhoods to co-design participatory methods centering Angelinos' understanding of energy access (Anderson et al., 2023; Romero-Lankao, Rosner, & Blanco, 2023; Rosner et al., 2023). It used a recognition-justice lens to examine how historical practices, such as zoning, renting, redlining, and discriminatory lending (Ong & González, 2019), have become embedded in current distributional energy inequities. For instance, previously redlined neighborhoods have higher-than-average summer temperatures, which have been found to lead to increased cooling loads and higher mortality risk during heat waves (Hoffman et al., 2020). Poor construction quality, and old, unsafe, and inefficient housing stock in underserved communities also limit equal installation of rooftop solar and EV charging infrastructure (Romero-Lankao et al., 2023a, 2023b). Angelenos have identified structural barriers, such as lack of financial means to purchase EVs or upgrade home electrical panels, that prevent them from benefiting from the green transition (Rosner et al., 2023).

Guided by listening sessions with underserved communities and steering-committee meetings with CBOs, the project evaluated 13 equity strategies (Romero-Lankao, Rosner, & Blanco, 2023). One strategy offers steppingstones to improve regulation, accountability, and enforcement of housing maintenance and upgrades and to reduce predatory practices among technology and service providers. Another strategy engages residents in developing programs that meet their needs, such as tailored training and professional development and resources on upgrading electrical panels and retrofitting buildings.

The city has also championed inclusive engagement in its just-transition planning by centering workers' voices in the transition away from fossil fuels. In 2022, the city government voted to end urban oil and gas drilling and established a Just-Transition Task Force. This task force, with representatives from the city and county, local communities, labor unions, academia, tribal nations, industry, and other key stakeholders has examined the demographic and economic/social effects on both unionized and non-unionized workers who might lose their jobs due to the phase-out of urban drilling. The task force has also taken up community and worker engagement and education programs, identified as a priority area based on surveys and interviews with oil workers. These findings also informed the city's Just-Transition Strategy, which supports workers in finding new employment, remediating decommissioned oil wells, and engaging front-line communities in land-use redevelopment planning (LA County & City of LA, 2022; C40 Cities, 2025).

Integrated Energy and Waste Solutions for Climate-Resilient Service Access in Kibera, Nairobi

Kibera, with an estimated population ranging between 200,000 and 700,000 people, is one of the largest slums in Africa (Desgroppes & Taupin, 2011). Despite its location near Kenya Power's existing electrical-distribution infrastructure, there are several barriers hindering residents' ability to access electricity formally. National grid supply in Kenya relies substantially on hydropower, and prolonged droughts in recent years have reduced output from major hydro reservoirs, forcing higher-cost imports and occasional rationing that typically deprioritizes informal settlements (Business Daily, 2025; EPRA, 2023). This has resulted in high upfront costs for grid connection, often amounting to three to five times the average monthly income of Kibera residents (de Bercegol & Monstadt, 2018). Moreover, once connected, consumers face additional charges, including value-added tax and fuel levies, further rendering electricity unaffordable for many. As a result, residents frequently turn to alternative energy sources such as charcoal and petroleum, or informal electricity connections through unskilled resellers, often without adequate regard for safety (de Bercegol & Monstadt, 2018).

The Human Needs Project (HNP), a US 501(c)(3) nonprofit, opened the Kibera Town Centre in 2014; the center (including its hybrid solar-microgrid) was developed and is operated by HNP using philanthropic contributions and grants raised by the organization. The microgrid currently serves a nearby school and a few local businesses, shielding them from the frequent power rationing (Francklyn, 2019). In addition, the power from the center is used to pump water from a borehole, providing clean water for the community. Complementary action by the Kounkuey Design Initiative has transformed frequently inundated river corridors into productive public spaces combining drainage, vegetation buffers, sanitation, and recycling micro-enterprises (Patel et al., 2022; WRI, 2021).

Coupling distributed energy with waste-sensitive, flood-resilient public spaces advances sufficiency and distributive justice; basic services are provided affordably, service continuity improves during shocks, and circular-waste activities create livelihoods.

How Barcelona Is Tackling Energy Poverty

The City of Barcelona is tackling energy poverty and reducing residents' vulnerability to climate change through an integrated, cross-sectoral approach driven by an understanding of energy poverty as a systemic issue that requires proactive policy.

Around 10% of Barcelona's population experiences energy poverty, and an additional 25% is at risk, particularly women with dependents, older adults, and migrants; this is exacerbated by intensifying heat waves that increase the need for cooling (Marí-Dell'Olmo et al., 2022). Recognizing this issue, the city made the eradication of energy poverty by 2030 a key pillar of its Climate Action Plan (City of Barcelona, 2018). In 2017, to protect the energy rights of residents and guarantee the application of Catalonian Law 24/2015[1] prohibiting companies from disconnecting vulnerable households from electricity and gas supply, the city created Energy Advisory Points (PAEs) in each district (City of Barcelona, 2025a). These advice points help people reduce energy costs and understand how to use energy effectively, and prevent companies from cutting off energy supplies. Since 2022, the portfolio of services provided by PAEs has expanded to include guidance on solar and green roofs, energy efficiency retrofits, and heat-related risk prevention.

Since inception, the PAEs have served over 222,000 people, corresponding to 87,350 homes, and its mediation efforts have prevented over 167,000 basic supply cuts and helped save 100 kW of electrical power every month (City of Barcelona, 2025b). To reach vulnerable populations, Barcelona also trained home-care workers to identify and address thermal discomfort in homes, especially among at-risk seniors (C40 Cities, 2024a, 2024b). The PAEs also promote green jobs by hiring long-term unemployed people and training them as energy advisors (C40 Cities, 2024a, 2024b). Every year the service hires 20 such residents, then

[1] Law 24/2015 is a regional law, representing an intermediate level of governance between the local and national government.

provides them with two months of training and 10 months of employment. Eighty percent of participants have successfully returned to the labor market after program completion.

Given the high demand and level of satisfaction among PAE service users, Barcelona has continued this municipal service and made it a core pillar of its climate-justice framework, and the city has better data on where energy poverty is experienced across the city and the impact on overall health and well-being (C40 Cities, 2022b; City of Barcelona, 2018; European Commission, n.d.).

Energy Access in Informal Settlements in Kampala, Uganda
In Kampala, a metropolitan area with a population of 6.46 million (UBOS, 2024), 60% of residents live in informal settlements (National Planning Authority, 2020). In Kisenyi, one of Kampala's largest informal settlements, 92% of households rely on polluting biomass fuels for cooking (ICLEI Africa, 2023), exacerbating health risks, gendered inequality, and deforestation.

Through the Enabling African Cities for Transformative Energy Access (ENACT) project, the Ugandan Ministry of Energy and Mineral Development, the Kampala Capital City Authority, and ICLEI have provided clean cooking solutions to over 10,000 residents (24% of the population) living in Kisenyi by deploying catalytic grant finance to clean cooking enterprises, helping residents to adapt technologies and scale business models that would otherwise be commercially unviable in underserved informal settlements. This approach strategically uses grant funding to de-risk innovation and lays the foundation for financially sustainable energy solutions. The Kampala City Energy Profile was developed to analyze household energy-use patterns and link this with government-led planning and infrastructure development. Policy recommendations on public-finance mechanisms for enabling market-led clean cooking access and analysis of financing mechanisms to support private-sector-led delivery models were developed to enable long-term financing of clean cooking in Uganda. Critical to these efforts is involving local SMEs as last-mile distributors, prioritizing affordability, flexible payment structures, and community-based after-sales services. To build public awareness and support behavioral change, a clean cooking video series and infographics were also produced.

Mexico City: Challenges in Equitable Water Governance Under Climate-Induced Scarcity

Since precolonial times, Mexico City has experienced water scarcity during the dry season. As illustrated by the severe drought in the summer of 2023, climate change is coalescing with urbanization and unsustainable water management to aggravate the risk of water scarcity. Usually, the rainy season starts in May or June, but spells of reduced rainfall, which have become more intense under climate change, make it difficult to replenish the water sources of a city that relies on overexploited underground aquifers for about 60% of its water needs. The rest is provided by external sources such as the depleted Cutzamala system, where water levels were so low in 2023 that city officials restricted water supply for several days at a time and raised concerns of a "Day Zero", when the system would be unable to supply any water to the city. For many decades the city has also been dependent on electrical systems to pump water from hundreds of meters underground and more than 100 km away to a city located 2200 meters above sea level (Romero-Lankao, 2010) highlighting the cross-jurisdictional, cascading implications of the need to supply water to a metropolitan area with over 22 million people.

Mexico City's enhanced water insecurity, due to overexploitation and systemic vulnerabilities in the system that are aggravated by climate change, is felt more acutely by households in informal and low-income settlements. During water emergencies, water-distribution trucks (also known as "pipas") cater to the needs of richer areas in the west, leaving poorer areas to suffer from shortages (Malkin, 2009), revealing inequalities in emergency water management.

To address some of these problems, Mexico City started the Water Harvest (Cosecha de Agua) program for populations facing water scarcity to employ nature-based solutions, such as urban reforestation and wetland restoration, which cool the city and improve water filtration, and to install rainwater-harvesting systems to improve access to water and reduce residents' reliance on the existing system. Between 2019 and 2021, more than 30,000 systems were installed, benefiting more than 150,000 people.

7.3 Processes: Inclusive and Just Decision-Making

7.3.1 *GRAA Pillars: Participation and Engagement, Informality, and Gender Justice*

Achieving sustainable and climate-resilient development requires the active participation of local communities, especially marginalized groups such as women, youth, migrants, slum dwellers, and informal workers, in shaping urban-development trajectories. Embedding community perspectives within formal decision-making processes facilitates the design and delivery of context-specific solutions that are more socially just and locally relevant.

Climate risks are unevenly distributed, yet structural inequalities are seldom reflected in climate policy. Local climate-policy frameworks must be designed to address and reduce structural vulnerabilities and existing social inequalities, informed by the situated knowledge, lived experiences, and specific needs of all urban residents, including those who are traditionally excluded from governance processes. A growing body of practice demonstrates how participatory approaches in climate action planning and implementation enhance legitimacy, accountability, and policy responsiveness (C40 Cities, 2020).

Across many Global South cities, more than one billion people reside in informal settlements that are highly vulnerable to climate hazards, reflecting an uneven spatial distribution of risk. However, they are also sites of emergent climate innovation and resilience led by local communities. This dynamic challenges formal planning paradigms that typically view informality as a constraint rather than a locus of creative adaptation.

Between 50 and 80% of informal settlement inhabitants are also employed in the informal sectors, which contribute significantly (between 25 and 50%) to non-agricultural GDP in low- and middle-income countries (ILO, 2018; Oduro-Appiah & Afful, 2020). In many cities, informal waste pickers collect 50%–100% of municipal waste, and informal transport systems may account for up to 90% of motorized trips in African contexts (Fair Circularity Initiative, 2024; WIEGO, 2023). Informal workers make indispensable contributions to urban climate mitigation and resilience, yet they endure systemic precarity, exposure to climate risks, unregulated work environments, and a lack of recognition, representation, and social protection.

Gender further shapes these occupational realities. Women often predominate in informal waste collection, street vending, and domestic services, yet they face additional barriers such as lack of safe public infrastructure, gendered mobility constraints, and exclusion from decision-making.

A growing number of cities are innovating participatory approaches that incorporate the lived experiences of informal workers (as in the case study on Accra below) and women (as in the case study in Panaji, Goa) into the planning and design of climate interventions, thereby enhancing equity, legitimacy, and effectiveness (C40 Cities, 2020; Fair Circularity Initiative, 2024a, 2024b).

Integrating informal innovation into formal climate strategies enables cities to make the most of low-cost, high-impact solutions for flood resilience, early warning systems, livelihood generation, and infrastructure upgrades, thereby enhancing equity, sustainability, and collective capacity in highly vulnerable urban contexts. Inclusive participatory approaches to climate action also recognize that informal settlement residents possess critical knowledge and agency, which, when harnessed through co-planning and participatory research, produce more effective and socially legitimate interventions; this is demonstrated in the case study on the Roof Over Our Heads initiative in India, described below.

7.3.2 *Case Studies*

Gender-Inclusive Streets in Panaji, India

Panaji, Goa—India's smallest state capital—is plagued with mobility challenges that restrict the movement of pedestrians: neighborhoods without footpaths, parking on spaces allocated for walking, and sprawl driven by motorways. The prioritization of private vehicles over pedestrians has undermined not only equitable access to city services and infrastructure but also efforts to meet the city's net-zero target for 2050.

To respond to these challenges, People's Urban Living Lab (PULL) initiated Inclusive Streets, a project designed to address gendered inequities and rising emissions in the city's mobility system (Bhandari et al., 2021). To understand the gendered dynamics of navigating the city by foot, PULL convened walks with diverse groups of women, such as vendors, students, tourists, commuters, and sanitation workers, each with unique relationships to the city's streets. These walks gave voice to the daily decisions women make to feel safe: commuters talked of leaving work early to avoid dark streets, and garbage collectors described how the absence of curb cuts and the presence of persistent obstructions on walking paths left them with no choice but to walk along busy roads, constantly at risk of collision.

Walking allows planners to "map, observe, and critically engage with urban spaces," revealing how design choices produce gendered experiences of accessibility, safety, and exclusion (Bassam, 2025). In contexts where streets serve not only as transit routes but also as places of livelihood, leisure, and social interaction, walking enables a deeper engagement with the complex social dynamics embedded in public space.

Inclusive Streets shaped a city-wide strategy with site-specific measures to improve women's mobility and support climate-friendly pedestrian infrastructure, which the city incorporated into projects to improve lighting and upgrade walking paths.

A key lesson from this project is that starting from how communities experience the city ensures that urban climate action delivers co-benefits that are equitable and publicly supported. Scaling such approaches requires embedding participatory methods into planning frameworks through sustained partnerships between local government and communities, and through capacity strengthening for planners and greater sensitivity to gendered challenges.

Engaging with Informal Workers to Improve Waste Collection in Accra, Ghana

In Accra, Ghana, the challenge of excessive waste and limited landfill space poses a severe threat to public health, with garbage contaminating waterways and toxic fumes from open-pit burning causing health issues and premature deaths, particularly for those residing near dumpsites. To meet the city's climate goals on waste, recycling, and methane-emissions reduction, as set out in its Climate Action Plan, the city identified the contribution of informal waste workers as critical and partnered with them to facilitate the sector's transition toward a green and just economy (City of Accra, 2020). The city held participatory stakeholder workshops that offered safe spaces for waste pickers to engage with waste-sector workers with knowledge of city bylaws and organizational capacity, and to put forward their concerns.

Through the engagement, informal workers expressed their needs for social protection and fair income, which the city is now addressing, notably access to social insurance, health care (extending coverage to up to 200 informal workers), and banking services. The city is also working with migrants in the city's informal waste economy to facilitate their access to financial support, health care, and childcare, and to strengthen their resilience to the impacts of climate change (C40 Cities, 2022c).

This concerted effort to enhance collaboration between the city and informal workers clearly shows how the three dimensions of justice—procedural, distributive, and recognition—intersect and foster the practical delivery of urban just transitions. In the last decade or so, the informal service providers have been able to increase collection coverage from 75 to 90%, waste capture from 53 to 90%, and recycling rates from 5 to 18%, saving the municipality US$5,460,000.00 in annual operational costs (Oduro-Appiah et al., 2019). The city's more recent engagement with informal waste workers has also led to a better understanding of their role in climate action and the development of just-transition policies that safeguard their livelihoods and ensure fair income, as well as resilience-building for urban communities against future climate threats (C40 Cities, 2022c).

Roof Over Our Heads Initiative: Surat and Pimpri, India

Roof Over Our Heads (ROOH) is both a program and a campaign focused on climate solutions for vulnerable communities, particularly those living in self-built, informal housing, that seeks to demonstrate the effectiveness of climate solutions rooted in their realities. ROOH emphasizes the importance of people understanding the logic and science behind housing choices so that they can identify solutions that are not only climate-resilient but also financially viable.

The campaign highlights the invisibility, illegality, and vulnerability of those without security of tenure, who nevertheless contribute significantly to the urban economy despite their precarious situation. ROOH seeks to expand their choices, enhance their resilience, and include them in decision-making at all levels. It calls on decision-makers to provide accessible finance, incorporate green materials, and include informal settlements in disaster risk management and climate-adaptation plans. By reframing informal residents as partners in climate survival, ROOH moves beyond traditional approaches to legalization, seeking to embed resilience into everyday living conditions and foster scientific innovation with practical development.

ROOH aims to establish 100 learning labs led by women's collectives and young professionals to co-produce housing solutions across 20 countries. The labs assess existing vulnerabilities and iteratively test solutions, returning to the drawing board if needed. Phase 1 involves a standardized assessment of both formal and informal materials, costs, and construction methods, entered into a shared database. In Phases 1 and 2, communities, contractors, material providers, and the ROOH team collaboratively design and build affordable, climate-resilient solutions, which are then tested for their ability to withstand environmental elements.

Starting from Indian cities like Pimpri, Ahmednagar, Surat, and Bhubaneswar, 50 ROOH Labs have been established across 10 countries, demonstrating how housing transitions can be incremental, co-produced, and climate-responsive. The labs facilitate the simplification of scientific knowledge, learning exchanges, and country-level roundtables that engage contractors, private-sector suppliers, and financial institutions to recognize low-income households as viable customers. The labs deploy a locally embedded woman-centric methodology that aligns community aspirations with technical feasibility, producing upgrades that are low-cost, thermally efficient, and governed by those who inhabit them. The initiative also collaborates with academic institutions globally to document and analyze housing-justice and climate-resilience efforts, and to integrate this into curricula.

The lives of families in the settlements of Azadnagar in Surat, India, were disrupted every monsoon season with water pouring into homes through cement corrugated roofs with layered tarpaulins that were inadequate for the intensity of the rain. Belongings were damaged or lost. School routines were interrupted. To address this, the ROOH Labs tested a range of affordable, climate-responsive materials, including fly-ash bricks, pre-painted galvanized zinc-coated roofing sheets for long-term durability and heat reflection, EPS insulation sheets to reduce heat transfer between walls shared between two houses, terracotta ceramic tiles for cooler flooring, and Vedic plaster, a breathable, natural wall finish that improves indoor air quality and reduces surface temperature.

With raised plinth and ceiling height, insulated walls, and breathable indoor plaster, the upgraded houses stayed dry. Not a drop of water entered. For the first time in years, belongings remained untouched and life continued indoors despite the urban flooding on streets. Improved ventilation also meant cooler indoor temperatures on hot days and a visible reduction in indoor smoke and soot. Fans that once ran 16 to 18 hours a day now run just three hours. Over months this has translated to a 25%–30% drop in electricity use. Families previously spent the equivalent of $22–$25 every year just to keep rain out, buying, layering, laying, and refastening tarpaulins repeatedly. Now they can use that money for other needs such as health care, education, and daily necessities. This work, co-produced with local skills and resident input, has further empowered women's collectives.

In Pimpri, in the state of Maharashtra, the local women's collective began questioning the materials and systems used in their houses due to concerns about high electricity bills, leaking roofs, and rising indoor temperatures. With technical support from the ROOH team, but decision-making power completely in their own hands, they adapted and implemented incremental upgrades. These included solar-reflective coatings, terracotta tiles, and insulated roofing sheets. The impact has been noticeable, as daily fan usage has dropped from around 12 hours to eight hours. Houses that once relied on seven tarpaulins for basic waterproofing now need only two. Managing flooding used to take up several hours a day and many sleepless nights, but with improved roofing and sealing, those efforts are no longer needed.

ROOH demonstrates that equitable climate transitions must begin with those most affected, as they are effective agents of change. When low-income women become co-creators of housing futures supported with the right tools, trust, and time, they generate systemic, scalable solutions that challenge the notion of informality as a barrier. Instead, it becomes a frontline of transformation. Resilience is not a luxury, but a necessity, and resilience in these towns is not aspirational: it is already happening. ROOH simply makes it visible, viable, and replicable.

Resilience From the Ground Up: Integrating Informal Systems in Beira's Climate Change Response

Beira, Mozambique, where over 70% of the population lives in informal settlements (Zhang et al., 2020), is highly vulnerable to flooding, tropical cyclones, and other climate-related disasters. In 2019, Cyclone Idai destroyed or severely damaged 90% of the city (Onishi, 2019). Despite extraordinary challenges, or perhaps because of them, Beira showcases how informality can be a site of innovation and resilience, not merely risk.

The Designing Inclusive African Coastal City Resilience (INACCT Resilience) project, led by ICLEI Africa, works to strengthen proactive, inclusive, and gender-responsive urban resilience planning and action in Beira. Building on the lessons learnt from Cyclone Idai, the project emphasizes how preparedness and collaboration can mitigate the scale of such devastation in the future. Central to this effort is the recognition of informal actors as knowledge holders and co-creators of just, locally informed climate-adaptation strategies with INACCT partners, embedded actors, and institutions (such as representatives of local committees for disaster risk management) working together as co-researchers to bridge the tacit knowledge of informal settlement residents with institutional planning and coordination.

Local committees for disaster risk management, volunteer-led, gender-balanced committees embedded in each informal settlement, are empowered by Mozambique's National Institute for Disaster Risk Management and Reduction to manage disaster preparedness and response at the local level. These committees coordinate evacuation drills, distribute aid, identify vulnerable households, and integrate local knowledge into emergency planning; these measures have proven instrumental in reducing disaster-related fatalities. Through learning labs, climate storylines, learning exchanges, ground-truthed mapping, community surveys, and photovoice methods, INACCT continues to support the critical integration of informal and formal systems to enhance a dynamic feedback loop between residents, researchers, and government entities, ultimately enabling meaningful action. In tandem with this inclusive governance structure, INACCT emphasizes that locally led innovations, such as water-retention ditches and bamboo reinforcements, are viable, community-driven strategies for bolstering climate resilience and informing citywide planning. Such insights emerging from Beira will contribute to the design of a knowledge-to-action platform that will support gender-responsive and inclusive planning for enhanced flood resilience across African cities.

7.4 Recognition: Addressing Structural Exclusions

7.4.1 Global Research Action Agenda (GRAA) Pillars: Intergenerational Equity/Youth Engagement; Indigenous Knowledge and Decoloniality; Crisis Response and Conflict

Recognition justice entails acknowledging and redressing the deep-rooted exclusions and structural inequities that shape how climate risks, responses, and resources are experienced across different urban populations. It goes beyond distributive fairness to confront the social, historical, and epistemic hierarchies that determine whose voices, knowledge, and identities are legitimized within urban climate governance (Fraser, 2000; Schlosberg, 2013). Within cities, recognition requires ensuring that groups historically marginalized, such as informal settlement dwellers, Indigenous peoples, migrants, and youth, are active agents defining what resilience and justice mean in their own contexts. As the climate crisis intensifies, recognition thus becomes the ethical foundation for equitable adaptation and transformative urban change.

As adaptation becomes untenable in some places, an estimated eight million people could move to just 10 cities by 2050 due to the climate crisis alone (C40 Cities, 2024a, 2024b). Cities, as the closest level of government to affected populations, are therefore critical actors in addressing loss and damage through proactive planning, inclusionary crisis response, and advocacy for local-level climate justice.

Recognition is also about epistemic justice: the fair valuation and integration of multiple knowledge systems in climate governance. The GRAA underscores that Indigenous and Local Knowledge and decolonial approaches are essential to ensure inclusive, locally driven adaptation (UN-Habitat, 2024). Yet, dominant climate research and urban-policy frameworks often remain rooted in colonial epistemologies that privilege technocratic and Western scientific paradigms over local, experiential, and Indigenous ways of knowing (Escobar, 2018; Whyte, 2017). Decolonization thus calls for dismantling these hierarchies by restoring the legitimacy of marginalized epistemologies and governance practices that have long sustained ecological balance and community well-being (Mignolo & Walsh, 2018).

Indigenous Knowledge systems offer relational understandings of human–nature interdependence, emphasizing reciprocity, stewardship, and collective responsibility (Berkes, 2018). When meaningfully integrated with scientific methods, Indigenous Knowledge enhances urban adaptation through context-sensitive environmental monitoring, early warning systems, and ecosystem restoration grounded in local ethics and values (Ford et al., 2016).

Recognition also requires intergenerational justice, acknowledging that the consequences of current actions extend far beyond present lifetimes. The inclusion of intergenerational equity and youth engagement in the updated GRAA UN-Habitat (2024) highlights that urban climate governance must account for the rights, needs, and agency of future generations. Historically, youth and children have been excluded from the decision-making processes that determine their collective futures, despite bearing disproportionate burdens from environmental degradation and unsustainable urban practices (UN-Habitat, 2024). Integrating intergenerational perspectives into planning ensures that today's adaptation and mitigation measures are resilient, forward-looking, and inclusive of those who will inhabit cities in decades to come.

This aligns with global policy momentum, such as the Declaration on Future Generations adopted at the 2024 Summit of the Future, which affirms the moral obligation of present generations to safeguard equitable, livable futures. Yet, the recognition of youth must move beyond symbolic inclusion toward genuine agency and co-production of climate solutions. Across the world, young people are demonstrating leadership in climate action through innovation, solidarity, and community-based initiatives. In Harare, the Young People's Federation of Zimbabwe is advancing circular-waste economies in informal settlements to mitigate flooding and pollution, while in Freetown, Sierra Leone, youth are restoring mangroves through the Freetown is Treetown campaign, using digital tools to track and monetize seedlings. These examples highlight youth not as passive beneficiaries but as co-creators of transformative urban resilience.

Many cities are increasingly demonstrating how recognition justice can reshape urban climate action by centering historically marginalized voices while integrating displaced and informal communities into planning processes. The following case studies from Mumbai, India, Cuenca, Ecuador, Atbara, Sudan, and Jigjiga, Ethiopia, illustrate how locally led and participatory approaches can advance more-inclusive and more-equitable urban futures. They show that when community knowledge, youth leadership, and displacement-affected populations are recognized as agents of change, climate solutions become more legitimate, contextually grounded, and sustainable.

7.4.2 *Case Studies*

Youth-Led Climate-Action Planning in Ambojwadi, Mumbai

Ambojwadi, a "non-notified" settlement in north-west Mumbai located one kilometer from the Arabian Sea, is excluded from official plans and denied basic services due to its unrecognized legal status. This exclusion deepens residents' vulnerability to flooding, heat stress, and other climate hazards, while preventing participation in formal planning processes. In collaboration with Youth for Unity and Voluntary Action (YUVA), a civil-society organization working to empower marginalized groups, local youth have taken the lead in documenting risks, advocating for infrastructure improvements, and driving equitable climate responses.

YUVA's Climate Justice initiative anchors climate resilience in community knowledge and agency. The organization mapped climate hazards and risks across Mumbai to identify "climate hotspots," where environmental stress intersects with urban poverty (Goswami & Parmar, 2024). Ambojwadi emerged as one such hotspot, characterized by poor housing, limited transport and sanitation infrastructure, and exposure to recurrent flooding.

YUVA has trained youth from various *pattas* (zones) to conduct vulnerability assessments combining climate and social data. As trusted residents, the youth co-designed surveys, conducted interviews, and mapped infrastructure gaps, generating grounded insights often missed by external actors. These efforts culminated in a Community Climate Action Plan (CCAP) outlining short-, medium-, and long-term adaptation priorities addressing both development and climate risks. The CCAP was followed by targeted advocacy and awareness campaigns with local officials and municipal representatives. Through sustained engagement, youth and community-based organizations secured municipal funding for drainage and sewage upgrades in several flood-prone *pattas* (Nuggehalli & Parmar, 2025).

The Ambojwadi CCAP demonstrates how locally generated data and youth leadership can shift climate governance from top-down planning toward co-created, place-based solutions. The initiative improved understanding of neighborhood-level vulnerabilities, empowered young people as intermediaries between communities and government, and unlocked tangible infrastructure investments that reduced flood risk and improved public health.

Beyond physical outcomes, the process strengthened civic capacity by embedding technical, advocacy, and organizational skills among youth participants, which in turn laid foundations for continued engagement with urban development and environmental justice. Ambojwadi's experience shows that equitable climate action must be grounded in recognition of informal settlements as legitimate urban spaces and of youth as key agents of change. It highlights how participatory processes, when incorporated into formal institutions, can align community aspirations with municipal priorities and enhance accountability in climate planning and that resilience is rooted in trust, participation, and recognition of marginalized voices.

This case study was produced based on an interview with Ms. Dulari Parmar, Project Lead—Climate Justiceat YUVA, India, in May 2025.

Sustainable Urban Integration of Displacement-Affected Communities in Atbara, Sudan, and Jigjiga, Ethiopia

Urban displacement driven by conflict, drought, and economic instability increasingly shapes African cities, creating overlapping vulnerabilities between internally displaced persons (IDPs), refugees, and host communities. The Sustainable Urban Integration of Displacement-Affected Communities (SUIDAC) program, funded by the EU and implemented by the *Cities Alliance*, operates across nine secondary cities in five African countries (DRC, Uganda, Ethiopia, Somalia, and Sudan).

SUIDAC promotes inclusive, rights-based urban development by improving physical infrastructure, fostering livelihoods, and strengthening governance mechanisms for equitable integration. The program focuses on displacement-affected communities often pushed to city peripheries, where overcrowding, poor services, and insecure tenure deepen marginalization. IDPs, refugees, host communities, and municipal authorities co-develop priorities for infrastructure, climate adaptation, and livelihood projects via local, national, and regional committees. This model ensures equitable participation and coordination across governance levels.

The program advances climate-sensitive urban planning, combining renewable energy and nature-based solutions to improve service delivery while reducing environmental degradation. It promotes green livelihoods, such as urban agriculture and small-scale eco-enterprises, linking economic resilience with environmental stewardship.

In **Atbara, Sudan**—where over 144,000 people have been displaced since 2023 due to armed conflict between the Rapid Support Forces and Sudanese Armed Forces—the municipality, Cities Alliance, and IOM are collaborating on inclusive planning for IDPs and vulnerable host communities. Efforts include using solar-powered systems to maintain water and health services during frequent power outages and training municipal staff in climate-smart planning. Livelihood programs support 70 micro- and cooperative industries involving about 3500 beneficiaries—mainly women, youth, and persons with disabilities—in green enterprises such as recycling and sustainable crafts. The integration of environmentally sensitive land use and flood mitigation into urban expansion plans will enhance the city's resilience to climate shocks.

In **Jigjiga, Ethiopia**, the influx of over 140,000 displacement-affected individuals fleeing drought and conflict has strained local infrastructure and resources. Here, interventions prioritize gender-sensitive sanitation facilities, school expansion, and vocational training tied to emerging green-job markets. Water and health infrastructure upgrades powered by solar energy improve access and resilience against service disruptions. The program reinforces local governance capacity and improved service delivery for marginalized populations while addressing both immediate humanitarian needs and long-term urban-development challenges. Inclusive sanitation and education projects reduce health risks and enhance access to opportunities for displaced populations. The program's participatory coordination structure, which links local communities, municipalities, and regional bodies such as IGAD, is aimed at improving cross-border policy coherence and climate-informed decision-making.

The Atbara and Jigjiga experiences underscore that climate innovation and social justice are mutually reinforcing. Solar-energy deployment, participatory infrastructure planning, and green-job creation show how decentralized, inclusive urban governance can deliver both climate resilience and social inclusion.

SUIDAC also demonstrates how community agency and local leadership can drive transformation even amid conflict and fragility. Its approach bridges humanitarian, development, and climate agendas, illustrating how cities can manage displacement not as a temporary crisis but as an opportunity for embedding equity into every stage of climate and urban planning, ensuring that those most affected by crises are central to defining solutions.

Integrating Indigenous and Local Knowledge and Representation into Water Management in Cuenca, Ecuador

Situated in the Andean highlands, Cuenca depends on the high-altitude páramo ecosystem for nearly all its water supply (Boelens, 2014). For centuries, Kañari communities have managed this high-altitude wetland through spiritual-ecological practices of reciprocity, communal labor, and respect for ecological thresholds (Boillat et al., 2013). Colonial expansion and modernization disrupted these governance systems through dispossession and privatization (de la Cadena, 2010). In recent efforts to restore Indigenous partnership, the municipal water utility, ETAPA, localized Ecuador's Fondo para la Protección del Agua model, channeling urban water tariffs into upstream conservation projects co-designed with Indigenous partners (Boelens et al., 2016). These initiatives integrate traditional practices such as native-species reforestation, soil regeneration, and the protection of sacred water sources into contemporary watershed management. Importantly, Indigenous representatives now hold formal decision-making roles in zoning, with nature-based solutions and climate programs marking a shift from tokenistic consultation to genuine power-sharing (Fricker, 2007; Santos, 2014). Cuenca thus exemplifies how decolonial urban climate governance can reconcile scientific and Indigenous epistemologies, embedding cultural ethics of reciprocity and sustainability within municipal systems.

7.5 Conclusion: Integration and Scaling

This chapter demonstrates that justice is a structural condition for successful urban climate action. The three justice dimensions (distribution, procedure, and recognition) are mutually reinforcing: distributive gains from improved service access rely on procedural inclusion to be fair and durable, and both depend on the recognition of local knowledge and identities to be legitimate and context-sensitive.

Several straightforward and actionable lessons emerge from the case studies outlined in the chapter. First, formal recognition of community and informal systems converts overlooked, local capacities into scalable civic infrastructure, improving reliability, lowering costs, and distributing benefits more equitably. Second, procedural justice requires sustained resourcing: genuine participation needs funded engagement processes, institutional pathways for uptake, and measurable accountability mechanisms. Third, equity-sensitive finance and metrics (targeted subsidies, sufficiency indicators, and disaggregated vulnerability data) are essential to avoid unintended harms such as displacement, and to ensure that benefits reach the most vulnerable. Fourth, empowering local governments with increased capacity and resources is critical for integrating community-led efforts at the city scale.

Scaling justice-driven innovation therefore requires three priority shifts: (1) policy and regulatory reform to legally recognize community efforts and utilities and informal-sector providers of basic and emergency services, so that municipal innovation is not blocked by outdated statutes; (2) dedicated finance (multi-year municipal capacity grants, catalytic subsidies for last-mile services) that follows locally defined priorities; and (3) monitoring frameworks that track process and equity outcomes, not merely aggregate emissions reductions: participatory data, disaggregated vulnerability mapping, and indicators for distributional outcomes must be core to climate-monitoring systems.

Finally, smaller and secondary cities, often overlooked, can be fertile testing grounds for rapid, community-embedded innovation and justice-oriented climate solutions, but capacity strengthening is a critical leverage point for scaling inclusive climate action in cities of all sizes. When justice is placed at the center of urban climate innovation, climate measures are more effective, more durable, and more likely to be socially legitimate. That is an ethical imperative and a pragmatic pathway to more-resilient cities.

Acknowledgements The authors would like to thank Aissatou Mbaye from Cities Alliance for her support to this chapter.

Competing Interests The authors have no conflicts of interest to declare that are relevant to the content of this chapter.

References

Anderson, K., Day, M., Romero-Lankao, P., Berdahl, S., Rauser, C., Bowen, T., Fournier, E. D., Heath, G., Hinojosa, R., & Ong, P. (2023). *LA100 equity strategies: Executive summary*. National Renewable Energy Laboratory (NREL).

Bassam, N. (2025) *Walking as a tool for change in the gendered city*. The Gendered City.

Berkes, F. (2018). *Sacred Ecology* (4th ed.). Routledge.

Bhandari, A., et al. (2021). Inclusive streets (Panaji, Goa): Gender-just design for non-motorized transport. *Journal of Urban Mobility and Equity, 3*(2), 45–61.

Boelens, R. (2014). Cultural politics and the hydrosocial cycle: Water power and identity in the Andean highlands. *Geoforum, 57*, 234–247. https://doi.org/10.1016/j.geoforum.2013.02.008

Boelens, R., Hoogesteger, J., & Rodríguez de Francisco, J. C. (2016). Commoditizing water territories: The clash between Andean water rights cultures and payment for environmental services policies. *Capitalism Nature Socialism, 27*(1), 84–102.

Boillat, S., Serrano, E., Rist, S., & Berkes, F. (2013). The importance of place names in Indigenous peoples' land management. *Human Ecology, 41*(4), 571–584.

Bulkeley, H., Edwards, G. A., & Fuller, S. (2014). Contesting climate justice in the city: Examining politics and practice in urban climate change experiments. *Global Environmental Change, 25*, 31–40.

Business Daily. (2025, May 28). *Optimum water levels in dams to power cheaper electricity*. Business Daily.

C40 Cities. (2019).*Executive guide: Inclusive planning*. https://www.c40knowledgehub.org/s/article/Inclusive-Planning-Executive-Guide?language=en_US

C40 Cities. (2020). *Resilience strategy for Accra: Integrating informal workers into climate action planning*. C40 Cities.

C40 Cities. (2022a). *Powering inclusive climate action in cities: Mapping city powers to deliver inclusive, just and equitable climate outcomes for all*. https://www.c40knowledgehub.org/s/article/Powering-inclusive-climate-action-in-cities?language=en_US

C40 Cities. (2022b). *City stories: Tackling energy poverty and building community resilience to crisis*. https://www.c40knowledgehub.org/s/article/City-stories-Tackling-energy-poverty-and-building-community-resilience-to-crisis?language=en_US

C40 Cities. (2022c). *Just transition applications in the informal waste sector.* https://c40.my.salesforce.com/sfc/p/#36000001Enhz/a/1Q000000giRY/.0d_.UD7YOjOYkNmGZuMXGkUBAajjOG3IbKRiAw0UdI

C40 Cities. (2024a). *Delivering a just energy transition in European cities through relief, retrofits and renewables: Resource pack for cities*. https://www.c40knowledgehub.org/s/article/Delivering-a-just-energy-transition-in-European-cities-through-relief-retrofits-and-renewables-Resource-pack-for-cities?language=en_US

C40 Cities. (2024b). *Inclusive climate action European Pilot: Barcelona.*

C40 Cities. (2025). *ICA fund: Collaborating with workers for a fossil fuel free LA*. https://www.c40.org/case-studies/ica-cities-fund-los-angeles/

City of Accra. (2020). *Accra climate action plan*. https://cdn.locomotive.works/sites/5ab410c8a2f42204838f797e/content_entry5ab410faa2f42204838f7990/5ab5605ea2f4220acf45cfa6/files/Accra_Climate_Action_Plan.pdf?1603293785

City of Barcelona. (2018). *Barcelona's climate action plan 2018–2030*. https://www.c40knowledgehub.org/s/article/Barcelona-s-Climate-Action-Plan-2018-2030?language=en_US

City of Barcelona. (2025a). *Energy advising points*. https://www.habitatge.barcelona/en/housing-services/problems-paying-your-home/energy-rights/energy-advising-points

City of Barcelona. (2025b). *Barcelona stops more than 167,000 basic supply cuts at its Energy Advice Points since 2016*. https://ajuntament.barcelona.cat/premsa/2024/10/15/barcelona-atura-mes-de-167-000-talls-de-subministraments-basics-als-seus-punts-dassessorament-energetic-des-del-2016/

De Bercegol, R., & Monstadt, J. (2018). The Kenya Slum electrification program: Local politics of electricity networks in Kibera. *Energy Research & Social Science, 41*, 249–258. https://doi.org/10.1016/j.erss.2018.04.007

de la Cadena, M. (2010). Indigenous cosmopolitics in the Andes: Conceptual reflections beyond "politics." *Cultural Anthropology, 25*(2), 334–370.

Desgroppes, A., & Taupin, S. (2011). Kibera: The biggest slum in Africa? *Les Cahiers D'afrique De L'est/The East African Review, 44*, 23–33. https://doi.org/10.4000/eastafrica.521

EPRA. (2023). *Energy and petroleum statistics report (2023)*. Energy and Petroleum Regulatory Authority.

Escobar, A. (2018). *Designs for the Pluriverse: Radical interdependence, autonomy, and the making of worlds*. Duke University Press.

European Commission. (n.d.). *Energy advice points (PAE)*.

Fair Circularity Initiative. (2024). *A living income for the informal waste sector*. Fair Circularity Initiative Report.

Francklyn, L. (2019). *Kibera microgrid helps bring education, clean laundry and employment to Nairobi's urban poor*.

Fraser, N. (2000). Rethinking recognition. *New Left Review, 3*, 107–120.

Fricker, M. (2007). *Epistemic injustice: Power and the ethics of knowing*. Oxford University Press.

Geels, F. W., Sovacool, B. K., Schwanen, T., & Sorrell, S. (2018). The socio-technical dynamics of low-carbon transitions. *Joule, 2*(11), 1821–1834.

Goswami, A., & Parmar, D. (2024). Youth-driven climate resilience and disaster preparedness: Grassroots lessons from Maharashtra. *Shelter, 25*(1), 66–78. https://www.ceew.in/sites/default/files/october-shelter-CEEW.pdf

Hoffman, J. S., Shandas, V., & Pendleton, N. (2020). The effects of historical housing policies on resident exposure to intra-urban heat: A study of 108 US urban areas. *Climate, 8*(12), 1–15. https://doi.org/10.3390/cli8010012

ICLEI Africa. (2023). *ENACT—Findings from a baseline survey on the status of clean cooking in Kisenyi, Kampala City in Uganda*. ICLEI Africa.

ILO. (2018). *Women and men in the informal economy: A statistical picture*. International Labour Organization.

IPCC. (2022). Cities, Settlements and Key Infrastructure. In *Climate change 2022: Impacts, adaptation and vulnerability*. Contribution of Working Group II to the Sixth Assessment Report. Cambridge University Press.

IPCC. (2025). *Scoping of the IPCC Seventh Assessment Report (AR7), cross-cutting topics: Consumption and equity*. IPCC-LXII/INF.7.

LA County & City of LA. (2022). *Los Angeles just transition strategy*. https://assets-us-01.kc-usercontent.com/0234f496-d2b7-00b6-17a4-b43e949b70a2/d2ade00b-66cc-4da1-8a01-7f9d72ee7b5d/LACounty-CityJustTransitionStrategy_FINAL12.5.22.pdf

Malkin, E. (2009). Mexico now enduring worst drought in years. *The New York Times*, August 12.

Marí-Dell'Olmo, M., Oliveras, L., Vergara-Hernández, C., Artazcoz, L., Borrell, C., Gotsens, M., Palència, L., López, M. J., & Martinez-Beneito, M. A. (2022). Geographical inequalities in energy poverty in a Mediterranean city: Using small-area Bayesian spatial models. *Energy Reports, 8*, 1249–1259. https://doi.org/10.1016/j.egyr.2021.12.025

Mignolo, W. D., & Walsh, C. E. (2018). *On decoloniality: Concepts, analytics, praxis*. Duke University Press.

National Planning Authority. (2020). *Third National Development Plan (NDPIII) 2020/21–2024/25*. National Planning Authority.

Nuggehalli, R. K., & Parmar, D. (2025). *Research: How local communities can assert voice in mapping*. Question of Cities.

O'Neill, D. W., Fanning, A. L., Lamb, W. F., & Steinberger, J. K. (2018). A good life for all within planetary boundaries. *Nature Sustainability, 1*, 88–95.

Oduro-Appiah, H., & Afful, K. (2020). Informal waste sector dynamics in urban Africa. *African Journal of Waste Management, 5*(1), 13–28.

Oduro-Appiah, K., Afful, A., Kotey, V. N., & de Vries, N. (2019). Working with the informal service chain as a locally appropriate strategy for sustainable modernization of municipal solid waste management systems in lower-middle income cities: Lessons from Accra, Ghana. *Resources, 8*(1), 12. https://doi.org/10.3390/resources8010012

Ong, P. M., & González, S. R. (2019). *Uneven urbanscape: Spatial structures and ethnoracial inequality*. Cambridge University Press.

Onishi, N. (2019). Flooding in Mozambique from Cyclone Idai made an 'inland ocean,' stalling rescues. *The New York Times*, March 19. https://www.nytimes.com/2019/03/19/world/africa/mozambique-cyclone.html

Patel, S. N., Ferrall, I. L., Khaingad, B., & Kammen, D. M. (2022). Sustainable and socially resilient minigrid franchise model for an urban informal settlement in Kenya. *Economics of Energy & Environmental Policy, 11*(1), 1–16. https://doi.org/10.5547/2160-5890.11.1.spat

Raworth, K. (2017). *Doughnut economics: Seven ways to think like a 21st-century economist*. Penguin Books.

Reckien, D., Juhola, S., Haque, A. N., et al. (2025). *Justice for resilient development in climate-stressed cities*. Cambridge University Press.

Romero-Lankao, P. (2010). Water in Mexico City: What will climate change bring to its history of water-related hazards and vulnerabilities? *Environment and Urbanization, 22*(1), 157–178.

Romero-Lankao, P. Rosner, N., & Blanco, L. (2023). Community-guided energy equity strategies. In *LA100 equity strategies*. NREL, LADWP, UCLA.

Romero-Lankao, P., Rosner, N., Brandtner, C., Rea, C., Mejia-Montero, A., Pilo, F., Dokshin, F., Castan-Broto, V., Burch, S., & Schnur, S. (2023a). A framework to center justice in energy transition innovations. *Nature Energy, 8*(11), 1192–1198.

Romero-Lankao, P., Rosner, N., Lockshin, J., & Zimny-Schmitt, D. (2023b). Justice as recognition. In *LA100 equity strategies*, 60.

Rosner, N., Blanco, L., Romero-Lankao, P., & Zimny-Schmitt, D. (2023). Chapter 2: Procedural justice. In *LA100 equity strategies*, 40.

Santos, B. de S. (2014). *Epistemologies of the south: Justice against epistemicide*. Routledge.

Schlosberg, D. (2007). *Defining environmental justice: Theories, movements, and nature*. Oxford University Press.

Schlosberg, D. (2013). Theorising environmental justice: The expanding sphere of a discourse. *Environmental Politics, 22*(1), 37–55. https://doi.org/10.1080/09644016.2013.755387

Spash, C. L., & Conway, E. (2021). Post-growth economics and sustainable consumption: Progress and challenge. *International Journal of Sustainable Development, 24*(2), 179–196.

Uganda Bureau of Statistics (UBOS). (2024). The national population and housing census 2024—final report—Volume 1 (Main), Kampala, Uganda.

UN-Habitat. (2024). *Accelerating next-generation city climate action: Findings from the 2024 Innovate4Cities Conference and update to the Global Research and Action Agenda on Cities and Climate Change Science (GRAA)*. UN-Habitat.

UN-Habitat. (2025). *UN-Habitat assembly opens with urgent global call to action on housing*. Nairobi, 29 May 2025.

Weindl, I., Schöll, L., & Gibon, T. (2022). Sufficiency for sustainability—Working definitions and implications for energy and material demand. *Resources, Conservation and Recycling, 183*, Article 106327.

Whyte, K. P. (2017). Indigenous climate change studies: Indigenizing futures, decolonizing the Anthropocene. *English Language Notes, 55*(1–2), 153–162.

WIEGO. (2023). *Informal workers and climate resilience: An evidence brief*. WIEGO.

World Resources Institute (WRI). (2021). *Nairobi public spaces build flood resilience*. https://www.wri.org/nairobi-public-spaces-build-flood-resilience%25utm_source%3Dchatgpt.%E3com

Zhang, X. Q., Eichenauer, I., Ryll, W., Segerlin, S., & Cannizzaro, N. (2020). *Financing for resilient and green urban solutions in Beira, Mozambique*. UN-Habitat.

Ziervogel, G., Pelling, M., Cartwright, A., Chu, E., Deshpande, T., Harris, L., et al. (2017). Inserting rights and justice into urban resilience: A focus on everyday risk. *Environment and Urbanization, 29*(1), 123–138.

https://www.c40.org/wp-content/uploads/2024/10/Digital-PDF_ICA-PILOT_Inclusive-Climate-Action-European-Pilot_final-23.09.24_02.pdf

CHAPTER 8

Digitalization for City-Level Models, Data, and Knowledge: Enabling Trust, Transparency, and Evidence-Based Climate Action

Pourya Salehi, David Corbett, Felix Creutzig, Dennis Pamlin, Simone Sandholz, Asad Asadzadeh, Zahra Assarkhaniki, Stelios Grafakos, Amy Jones, Pooja Mahapatra, Juan Pablo Astolfo, Valentina Palermo, Giulia Melica, Martin Wainsten, Angel Hsu, and Theo Kötter

8.1 Introduction

By 2050, nearly 70% of people will live in cities that already account for more than 70% of global greenhouse gas emissions (Mukim et al., 2023). Urban areas, therefore, concentrate both climate risks and the capacity to

P. Salehi (✉) · A. Jones
ICLEI—Local Governments for Sustainability, Bonn, Germany
e-mail: pourya.salehi@iclei.org

A. Jones
e-mail: amy.jones@solutionzone.co.uk

C. Oke et al. (eds.), *A Global Research and Action Agenda on Cities and Climate Change Science*,
https://doi.org/10.1007/978-981-95-6517-7_8

deliver solutions, from innovation and finance to civic engagement (Yan & Cheng, 2024). The effective use of city-level models, data, and knowledge is now central to closing the gap between ambition and implementation.

D. Corbett
ClimateView, Stockholm, Sweden
e-mail: david@climateview.global

F. Creutzig
Potsdam Institute for Climate Impact Research, Potsdam, Germany
e-mail: creutzig@pik-potsdam.de

D. Pamlin
RISE Research Institute of Sweden, Gothenburg, Sweden
e-mail: dennis.pamlin@ri.se

S. Sandholz
Institute for Environment and Human Security, United Nations University, Cologne, Germany
e-mail: Sandholz@ehs.unu.edu

A. Asadzadeh
Institute of Geodesy and Geoinformation, University of Bonn, Bonn, Germany
e-mail: asad.asadzadeh@uni-bonn.de

Z. Assarkhaniki
Faculty of Spatial Sciences, University of Groningen, Groningen, The Netherlands
e-mail: z.assarkhaniki@rug.nl

S. Grafakos
Global Green Growth Institute, Budapest, Hungary
e-mail: stelios.grafakos@gggi.org

P. Mahapatra
Fugro, The Randstad, The Netherlands
e-mail: p.mahapatra@fugro.com

J. P. Astolfo
Global Covenant of Mayors, Brussels, Belgium
e-mail: jpastolfo@globalcovenantofmayors.org

V. Palermo · G. Melica
European Commission, Joint Research Centre (JRC), Ispra, Italy
e-mail: Valentina.PALERMO@ec.europa.eu

Since 2018, the Global Research and Action Agenda on Cities and Climate Change Science (GRAA) and successive Innovate4Cities (I4C) cycles have clarified what cities need from digitalization: actionable data that is trustworthy, transparent, and usable across institutions. The updated agenda organizes these needs into eight linked dimensions, or towers, that guide how information should be generated, governed, and applied. This chapter adopts those dimensions as a practical lens for strengthening city data ecosystems.

We define digitalization as the full integration of digital tools, data streams, and analytics into urban governance. City knowledge encompasses baseline inventories, scenario and risk models, real-time sensors, citizen-generated inputs, and place-based expertise that guide local decision-making. A systems approach recognizes feedback cycles, trade-offs, and path dependencies across all levels (Salehi & Sharifi, 2022). The aim is to align system demand trajectories with delivery readiness so that data translates into investable action.

Three developments motivate the analysis. First, cities are moving beyond static inventories toward longitudinal, interoperable data infrastructures that connect mitigation, adaptation, and equity. Second, digital tools such as artificial intelligence (AI), internet of things (IoT), and digital twins can shorten decision cycles when paired with clear governance and capacity. Third, investors and treasuries increasingly require transparent, reproducible models that clarify how interventions deliver outcomes over time.

G. Melica
e-mail: Giulia.MELICA@ec.europa.eu

M. Wainsten
OpenEarth Foundation, San Diego, CA, USA
e-mail: martin@openearth.org

A. Hsu
University of North Carolina at Chapel Hill, Chapel Hill, NC, USA
e-mail: angel.hsu@unc.edu

T. Kötter
University of Bonn, Bonn, Germany
e-mail: tkoetter@uni-bonn.de

The chapter contributes in three ways. It charts how city knowledge systems have evolved, highlighting shifts in goals, roles, and technologies that change how municipalities plan and deliver. It brings to the surface persistent challenges, including interoperability gaps, algorithmic risks, uneven capacity, and the environmental footprint of digital systems. It then presents practical responses through three case studies and a systems-thinking synthesis that links open standards, participatory data, ethics-by-design, and finance readiness.

We first outline the evolution of city-level data and knowledge systems. We then use the eight GRAA towers to summarize key challenges. Next, we examine how digitalization and capacity-building enable credible, scalable responses. Three case studies illustrate implementation at city and country scales. A systems-thinking section connects the towers to considerations of delivery and justice. We close with a concise set of recommendations and metrics for governments, financiers, and civil society.

8.2 The Evolution of City-Level Data and Knowledge Systems

Since the early 2000s, cities have adopted data-driven methods for climate action, but the purpose and use of such data have changed markedly since the first GRAA in 2018. Early efforts focused mainly on mitigation, especially on greenhouse gas inventories based on the Global Protocol for Community-Scale Greenhouse-Gas Emission Inventories (GPC) (ICLEI, C40, & WRI, 2014). These provided annual snapshots of emissions but were rarely linked to adaptation, resilience, or equity (Bulkeley et al., 2014). The 2018 conference that produced the first GRAA expanded the agenda to include systemic interdependencies among infrastructure, governance, and climate risk (Hunter et al., 2023). Local governments evolved from data implementers to co-producers, using evidence to guide land-use, finance, and infrastructure choices (Taylor et al., 2021). Data became a strategic asset for decision-making rather than a compliance requirement.

Methods also shifted from linear, supply-side accounting toward dynamic, systems-based approaches. Cities now increasingly view mitigation and adaptation capacity through their ability to generate, connect, and apply knowledge across sectors (Hsu et al., 2019). Participatory sensing, open-data portals, and community engagement allow citizen-generated data on air quality, flooding, and urban heat to complement official statistics (West et al., 2019). When such "smart" systems are transparent and inclusive, they enhance legitimacy and uptake (Jackson et al., 2022).

The virtual 2021 Innovate4Cities Conference saw the key cross-cutting issue of *City-level models and data* joined by *digitalization* as a dimension of consideration (this is discussed in detail in Chapter 3). These dimensions defined the structural qualities of actionable, equitable data systems, and emphasized interoperability and trust over technical sophistication alone. Digital tools moved from optional pilots to essential infrastructure for traceability and accountability (Batty, 2018). Urban digital twins, combining real-time sensors with AI-based simulations, let planners test interventions virtually before implementation, mainstreaming model-based governance (Law & Zhang, 2018; van Berlo et al., 2020). As the concept has spread, authors have suggested practical ways to differentiate digital twins based on factors like data exchange between physical and digital systems, system level, and cognitive features, with interoperability and life-cycle integration often lacking in real-world applications (Mendonça et al., 2024).

The 2024 Innovate4Cities conference advanced this agenda by formalizing eight "towers" of city-level data, models, and knowledge: communication, decentralization, digitalization, efficiency, information integrity and transparency, risk, scale, and uncertainty—that reframed city climate knowledge as a shared commons co-created among governments, citizens, and researchers (Irvin et al., 2024). This shift rejected techno-solutionism and stressed that trust, transparency, and governance capacity determine digital success (Huang-Lachmann, 2019).

Recent initiatives illustrate this evolution. The Swedish ClimateView platform enables cities to link system-level demand (e.g., modal-shift targets) with interventions and forecasted outcomes, aligning strategic vision with delivery (Shalit et al., 2024). In Brazil, DUCTExplorer applies AI and spatial analytics to map heat vulnerability and collect community feedback, while the CityCatalyst stack translates national datasets into city-scale assessments. These examples show the shift from static reporting toward decision-supporting systems.

Another significant change is the convergence between digital knowledge systems and climate finance. Transparent, reproducible models that show how interventions deliver results are becoming prerequisites for investment, connecting planners, financiers, and donors (Hughes et al., 2020; Shalit et al., 2025). As competition for resources grows, continuous monitoring of outcomes across sectors strengthens credibility and accelerates project approval.

Overall, the past decade has marked a transformation in urban climate governance. Cities act as strategic data stewards, designing and operating infrastructures that connect global goals with local action. From the early greenhouse gas inventories of the 2000s to today's AI-enabled digital twins and participatory platforms, digitalization has reshaped both technology and institutions, paving the way for more transparent and finance-ready transitions.

8.3 Key Challenges in City-Level Models and Data

Despite progress, major gaps persist that limit the effectiveness of urban digital systems.

8.3.1 Data Availability and Interoperability

High-quality, standardized city-scale climate data remains scarce. Many local governments -especially in the Global South- still make critical decisions with incomplete datasets, while even data-rich cities suffer from fragmentation across departments and vendors (Yang et al., 2024; Bozkurt et al., 2025). The 2021 Innovate4Cities report ranked "observations, data and modelling at the city scale" among the least developed areas of the GRAA (GCoM & UN-Habitat, 2022).

8.3.2 Capacity and Continuity

Knowledge systems depend on expertise, funding, and institutional memory that often change with election cycles and donor programs. Evidence from the 2021 Western Germany floods further illustrates this challenge in practice: disaster response across interconnected water and health systems was limited by gaps in coordination, communication, information management, and timing, worsened by physical disruptions and the legal and policy environment, leading to inadequate preparedness and response (Nick et al., 2023). Additionally, when pilots end or trained staff leave, monitoring collapses. The Tuvalu Coastal Adaptation Project, funded by the Green Climate Fund, delivered the country's first high-resolution elevation model (UNDP, 2021), but it also illustrates how valuable datasets can degrade without long-term stewardship.

8.3.3 *Standards, Transparency, and Algorithmic Accountability*

AI-based decision tools offer analytical power, yet risk bias, and opacity. Proprietary optimization systems have disadvantaged informal settlements when training data have omitted them (Sanchez et al., 2024). Without audit protocols, cities may rely on "black-box" tools that erode trust. In safety-critical fields such as disaster risk management, practitioner guidance stresses that bias, privacy and security risks, lack of transparency and explainability, and inflated expectations can cause real harm if machine learning tools are used without safeguards and governance mechanisms (Soden et al., 2021). Open, modular frameworks such as ClimateView's Transition Elements (Shalit et al., 2024) demonstrate how transparent schemas can standardize assumptions and improve reproducibility.

8.3.4 *Risk, Uncertainty, and Downscaling*

Translating national or global models to neighborhood scales is complex. Simplifying land-use or socio-economic patterns can cause maladaptation (Wilby et al., 2014). Many cities are shifting to opportunity-based scenario planning that co-creates desirable futures, backcasting to identify enabling actions (Haasnoot et al., 2013). These approaches require participatory tools and a cultural shift from risk avoidance to adaptive learning.

8.3.5 *Decentralization Versus Efficiency*

Distributed sensors and citizen science enhance local insight but raise questions about data quality and validation. Studies show that wearable air-quality sensors can achieve reference-grade accuracy through clustering and calibration (Nguyen et al., 2025). Yet such methods can exceed municipal capacity, while vendor-locked centralized platforms restrict flexibility. Balanced governance should encourage experimentation within interoperable standards.

8.3.6 Environmental Footprint of Digitalization

Running complex AI models and digital twins consumes significant energy; AI data centers could reach 3.4% of global electricity demand by 2030 (de Vries, 2023). Hardware production also carries embodied carbon and resource costs (Zewe, 2025). Quantifying and offsetting this footprint is crucial to ensure that digital tools support, rather than undermine, mitigation goals.

8.3.7 Data Ethics and Epistemic Justice

Informal, Local, and Indigenous Knowledge remain under-represented in formal datasets. A review of planetary-health citizen-science projects found that institutions in the Global North funded 87% of projects, potentially skewing priorities (Ambole et al., 2025). Recognizing diverse epistemologies prevents inequities and enriches understanding of place-specific resilience.

These challenges are institutional, ethical, and technical. Interoperability without transparency can reproduce inequality, and hidden carbon costs may offset efficiency gains. Operating the eight GRAA towers using mutually reinforcing principles—openness, accountability, fairness, and continuity—will allow cities to transform fragmented initiatives into inclusive, enduring knowledge systems.

8.4 Digitalization and Capacity-Building for City Climate Action

Digitalization has moved to the center of urban climate governance. AI optimizes transport routes in real time, IoT sensors stream micro-climate data, and digital twins allow policy scenarios to be tested before implementation (Argyroudis et al., 2022; Helmi et al., 2024). When used strategically, such tools shorten decision cycles and build the evidence base that financiers need. Their success, however, depends on human and institutional capacities that remain uneven.

AI analytics now turn raw sensor data into hourly emissions baselines, enabling carbon budgets to be managed like ledgers (Shen et al., 2023). IoT networks track heat islands, groundwater, and building energy use, feeding early-warning dashboards for adaptation (Balogun et al., 2020). Continuous digital-twin operation can deliver up to 18% energy savings in retrofit programs (Bortolini et al., 2022). These systems transform episodic inventories into living knowledge platforms.

However, technical innovation must be matched by skills and governance. Many municipalities lack personnel with statistical or coding expertise; audits show that fewer than one in 10 climate-staff positions in African and Asian cities require such skills (Bastidas et al., 2024). Broader training frameworks like DC2-CF combine data competence with systems thinking, stakeholder engagement, and ethics-by-design (Helmi et al., 2024). Roles should be defined by function: compliance analysts need methodological rigor, while transition modelers need facilitation and prototyping abilities.

Financing remains a major constraint. Sensors, cloud services, and expert staff entail high upfront costs. Positive net benefits emerge only when technology upgrades coincide with governance reforms that streamline project pipelines and enable blended finance (OECD, World Bank, & UN Environment, 2018). Mechanisms such as outcome-linked bonds or pooled procurement can improve affordability. The WHO Small Island Developing States initiative showed how harmonized procurement yielded average savings of 56% (World Health Organization [WHO] Regional Office for Africa, 2024).

Digital platforms also connect fragmented project pipelines with climate-aligned capital. When data demonstrates system-wide interactions, such as e-buses cutting CO_2 while freeing grid capacity, treasuries can present program-level business cases that align with investors' risk appetites. Transition portfolios combining multiple measures under shared metrics can reduce transaction costs by 30% and attract larger investments (Monasterolo & Battiston, 2022; Shalit et al., 2025).

Ethics and trust are central. Public–private data-trust models pool municipal and citizen data under neutral charters guaranteeing anonymity, open-access derivatives, and transparent governance (Verhulst, 2024). Participatory ethics boards enable civil society to review algorithms and address privacy and surveillance risks (Sanchez et al., 2024). Such safeguards strengthen legitimacy and encourage long-term data sharing.

The Human Needs-Based Climate Innovation Framework (UNFCCC, 2021) reinforces people-centered evaluation by measuring technology through outcomes such as clean energy, safe mobility, and resilient shelter, rather than hardware counts. Linking each digital indicator to a needs-based result translates technical progress into social benefit. RiskScape, for example, prioritizes flood-mitigation investments for residents of informal settlements, aligning adaptation with equity.

Four enabling conditions support responsible digitalization:

1. *Open standards* that let cities integrate new datasets without re-engineering (e.g., Transition Element Framework, GPC).
2. *Outcome-based finance* that ties funding to independently verified data.
3. *Ethics-by-design training* covering privacy, bias auditing, and life-cycle carbon accounting.
4. *Interoperable governance* via data trusts and citizen oversight that balance innovation with accountability.

With these capabilities in place, cities can move from pilots to adaptive, scalable, and socially responsive digital climate practice.

8.5 Case-Study Examples: Digital Strategies for Climate-Responsive Cities

The rapid development of digital tools has transformed how cities understand, plan, and monitor climate action. While early climate tools mainly focused on creating emissions inventories or tracking individual projects, today's platforms are increasingly designed to support systemic planning, integrate with financial frameworks, and enable ongoing assessment. At the same time, new forms of digital infrastructure, such as participatory mapping and interoperable data standards, are allowing cities to transition from single assessments to adaptive, evidence-based decision-making. This section presents three examples of digital tools used to support climate-conscious urban governance. Each example follows a standard structure: context and motivation, approach and features, emerging impacts, and lessons learned. Collectively, these cases demonstrate how cities and their partners are developing the next generation of digital climate strategies. After the case studies, we examine the growing role of co-production in shaping city knowledge systems, including a snapshot from Tuvalu's coastal-resilience initiative.

8.5.1 Case Study 1: ClimateView

Context and Motivation

Cities are facing growing pressure not only to demonstrate that their climate strategies yield tangible results but also to use data to drive and inform their decisions. Traditional methods often depend on emissions inventories and sector-specific plans that are disconnected from actual implementation or investment plans. The gap between strategic goals and operating capacity has created a need for specialized platforms that take a holistic approach, aligning data, roles, funding, and monitoring (Lonergan et al., 2025). The ClimateView platform, now adopted in more than 250 cities worldwide, was developed to address this need, helping cities structure and model their climate transition as a series of activity shifts based on IPCC mitigation options, monitor implementation readiness, build investment confidence, and share knowledge with other cities.

Approach and Features

The architecture of ClimateView takes a systems-based approach to structuring transition efforts by defining goals, understanding the current status, identifying needed shifts in activities and their requirements to occur, developing, and logging interventions to fulfill the requirements to achieve those shifts, and using custom indicators to monitor change and inform iteration. The method is grounded in the Transition Element Framework (TEF), which provides a standardized approach (DIN SPEC 91637[1]) for defining, comparing, and monitoring activity shifts across sectors, including transportation, energy, and housing. Using this framework, cities can collaborate more effectively to develop and implement evidence-based transition pathways grounded in IPCC science and tailored to local constraints and opportunities (Colona, 2023).

The platform supports the operationalization of systems thinking and multi-level governance. For example, in Scotland, the Scottish Climate Intelligence Service (SCIS) uses ClimateView to help all 32 local authorities coordinate their climate plans while maintaining local context. In Germany's Ruhr region, 53 municipalities are digitizing local interventions that, together, form a regional scenario, enabling spatial-planning

[1] https://www.din.de/en/innovation-and-research/din-spec-en/current-din-specs/wdc-beuth:din21:393817040.

agencies to identify gaps, align actions, and make a case for state-level support. The platform features live dashboards, intervention mapping, and integration with reporting standards.

Emerging Impacts

Results indicate improvements in both internal and external coordination. Cities and regions using ClimateView report better alignment between departments, stakeholders, and levels of government (Scottish Climate Intelligence Service, n.d.). From a finance perspective, ClimateView supports clear monitoring and communication of impact, as well as the aggregation of smaller actions into portfolio-level investment cases. For example, cities in the Netherlands are using ClimateView to standardize their housing and mobility plans under a national data playbook, resulting in comparable outputs that may enhance investor confidence.

Lessons Learned

ClimateView shows the importance of combining scientific accuracy with planning usability. Cities need tools that not only track emissions but also map interdependencies, measure system demands, and show how they achieve results, unlocking innovative financing options and enabling implementation. Instead of having an isolated impact, ClimateView enables cities to navigate their entire transition through a comprehensive, integrated approach. The platform proves that when system-wide readiness indicators and co-benefit tracking are integrated into a single digital environment, it becomes easier for cities to secure funding, coordinate stakeholders, and adjust their plans based on real-world feedback. Therefore, ClimateView functions as an operational infrastructure for multi-level governance and climate action.

8.5.2 Case Study 2: Brazil's CHAMP Pilot and the CityCatalyst Digital Stack

Context and Motivation

Brazil's CHAMP pilot (formally titled *"Scaling High-Impact Mitigation and Adaptation Actions in Brazilian Cities on the Road to COP30"*) is the first large-scale digital demonstrator under the Bloomberg Philanthropies Joint Program led by C40, the Global Covenant of Mayors (GCoM), and CDP. Conceived as a flagship implementation of the Coalition of High Ambition Multilevel Partnership (CHAMP), the project demonstrates how a country can translate abundant national-level climate datasets into actionable city-level strategies at speed and low cost.

Brazil was chosen for three strategic reasons. First, it hosts more than 5500 municipalities, giving CHAMP an unparalleled test bed. Second, the federal government has already signed the CHAMP pledge, making a commitment to embed sub-national action in its 2025 NDC. Third, Brazil maintains some of the world's richest open datasets, including SEEG for greenhouse gas estimates, AdaptaBrasil for hazard rasters, and IBGE for socioeconomic statistics. However, most city halls lack the tools and bandwidth to convert these data streams into greenhouse gas inventories, climate-risk profiles, and bankable projects.

To close that gap, the consortium deployed CityCatalyst, an open-source platform developed by the Open Earth Foundation. This AI-enabled platform ingests national datasets through modular APIs, cleans and harmonizes them in a central data framework, and serves role-based dashboards to municipal officers, state secretariats, and federal ministries. The same data spine powers three automated assessments: (i) a baseline greenhouse gas inventory, (ii) a climate change risk assessment, and (iii) a high-impact-action scoring engine that ranks mitigation and adaptation options by abatement potential, resilience benefit, equity impact, and legal authority. This was initially developed for 50 cities, and has recently been expanded to all 5570 municipalities under Brazil's Green Resilient City Program.

Approach and Features

The implementation of CityCatalyst follows a six-step digital playbook codified in the project's methodology: multi-scale coordination to create a shared evidence base; a data-driven assessment combining greenhouse gas inventories and climate change risk assessment that is generated in weeks rather than months; prioritization of the highest-leverage actions using an AI-powered "tournament ranking between actions and the city profile"; the use of generative AI to develop initial implementation plans for each action; matching those actions with finance opportunities; and the establishment of MER indicators that flow automatically to the CDP-ICLEI Track in line with the GCoM Common Reporting Framework. Finally, engagement workshops delivered by local consultants enable cities to iteratively refine their data, providing a template for replication across other CHAMP countries.

Emerging Impacts

Early results from the first five pilot cities underline the digital dividend:

- Cost reduction: Approximately US$4,800 external cost per city, compared to historical deep dives of US$25–35,000.
- Time reduction: less than six weeks from kick-off to validated GHGI + CCRA, compared with six to 12 months using spreadsheet templates.
- Good enough data to "get started": 60 inventory line items auto-filled from national datasets; remaining gaps captured through a two-week in-app survey.
- Action acceleration: The identification of 100 high-impact actions (one mitigation and one adaptation per city) enables cities to launch these as initiatives and create shared economy-of-scale opportunities at the national level.

Lessons Learned

The pilot exposed frictions, chiefly in the form of limited access to disaggregated utility data and the need to coach smaller municipalities on interpreting dashboards; however, these issues are being addressed through a 2025–2026 training program delivered by local consultants. Another aspect is that while working with 50 cities simultaneously, "one size doesn't always fit all," as cities have diverse capabilities and needs, requiring dedicated human interpretation to adapt findings to each context. As an initial, extensive application of AI and data pipelines, the project and tool have proven successful, but with the advent of AI technology, the potential to improve action prioritization and turn specific actions into bankable projects is expected.

In sum, the Brazil pilot demonstrates that when national open data, modern digital architecture, and multi-level governance are braided together, city climate transition can shift from bespoke consultancy exercises to a repeatable public-good service, accelerating the journey from political commitment to measurable, monitored progress on the ground.

8.5.3 Case Study 3: DUCTExplorer

Context and Motivation

The 2024 Innovate4Cities Conference (I4C24) created a high-impact space for exploring how AI can support urban climate action. As part of this effort, the AI x City Climate Action Hackathon invited participants to design AI-driven tools for assessing climate risks in Brazilian cities, including extreme heat, floods, sea-level rise, and droughts. The DUCTExplorer team from the Singapore-ETH Centre, led by Adelia Ayu Sukma, was chosen as the winner, and was matched with the city of Belo Horizonte to pilot their solution in the Brazilian context.

Approach and Features

The DUCTExplorer team developed a flexible and scalable AI framework to support city-level climate vulnerability assessments. To overcome the challenge of mapping urban heat, they proposed using a Digital Urban Climate Twin (DUCT). This approach integrates machine-learning-enhanced data sources with physics-based climate models to create a comprehensive representation of urban climates across diverse scenarios, helping decision-makers identify the urban areas most vulnerable to heat exposure.

The methodology combines machine learning, GIS analysis, and open data (e.g., WorldPop, IBGE, INMET) to produce city- and neighborhood-level risk assessments. Users can interact with the system via a dashboard that visualizes population exposure, climate stressors, and infrastructural risks. The platform was designed to be transparent, modular, and replicable, enabling application in other CHAMP countries with minimal adjustment.

Emerging Impacts

The Belo Horizonte pilot has demonstrated the tool's great adaptability to diverse natural (e.g., height, wind) and socio-economic conditions. For specific hazards such as heat, cities need the capacity to analyze, with granularity, the different levels of exposure faced by the most vulnerable population groups, and to explore the impact of actions. City officials in this Brazilian city were highly engaged with the model's results, which provided evidence to support their assumptions about risks to their population and enabled them to analyze potential solutions, such as increased linear urban trees, green spaces, and green corridors.

Lessons Learned
DUCTExplorer exemplifies how AI can bridge the gap between climate data and actionable policy. This case highlights the value of combining academic expertise, civic data, and user-friendly design. The Innovate4Cities Hackathon showcased how interdisciplinary collaboration can rapidly yield tools with real-world utility that cities need now more than ever. Tools like DUCTExplorer have the potential to inform climate-adaptation strategies, helping cities turn risk into resilience through informed, data-driven action.

8.5.4 Discussion

The increasing use of digital technologies has changed the way data is created, shared, and used in cities. Local governments now have access to sensor networks, social media, and machine-learning tools to produce and analyze large amounts of data (Lember et al., 2019). In this context, the knowledge held and created by citizens about how they use and live in the city is seen as highly valuable (Ataman et al., 2025). Therefore, co-production and co-generation of knowledge become crucial.

The proliferation of wearable and sensing technologies enables the collection of real-time, detailed data across various aspects of urban life, opening new opportunities for citizen participation and contribution in data collection. Closely related to broader citizen science initiatives, crowdsourcing has significantly altered data-collection methods by enabling information to be gathered from multiple people working together toward a common goal (Gebremedhin et al., 2020; Haklay, 2012). Volunteered geographic information enables non-experts to create and share location-based data (Goodchild, 2007). In contrast, community-based monitoring allows local groups to systematically track environmental and urban issues over time.

However, increased reliance on user-driven data also presents challenges, such as the need for clear accountability, data-quality assurance, and seamless system integration. Despite these concerns, studies have shown that citizen-generated data can meet quality standards comparable to those of official sources (Ponti & Craglia, 2020). Participatory processes, ranging from focus groups to participatory GIS, are increasingly supported by digital platforms (de Vos et al., 2021), thus enhancing legitimacy, inclusiveness, and behavioral change.

The Tuvalu Coastal Adaptation Project, supported by the Green Climate Fund and UNDP, shows how digital data systems can enhance climate resilience in small island developing states. High-resolution aerial surveys have delivered unprecedented elevation and topographic data across Tuvalu's islands, which have been used to redesign roads, guide infrastructure projects, and evaluate coastal vulnerability. However, while technically strong, the long-term impact of this data depends on institutional capacity to interpret, manage, and use it effectively. The Tuvalu example underscores the importance of integrating digital tools with investments in data stewardship, literacy, and governance to enhance local planning and resilience.

As cities face increasingly complex climate challenges, digital platforms and participatory data systems provide new ways to organize and implement transitions. The ClimateView example shows how a systems-based, finance-aligned approach can turn strategy into action. Future examples from OpenEarth and DUCTExplorer will add more depth to this picture. Ultimately, co-produced data and digital infrastructure must work together to build the knowledge ecosystems that cities need for transformative, inclusive, and accountable climate action.

8.6 A Systems-Thinking Approach to Digitalization and Climate Action

Digitalization increasingly functions as a city's nervous system, transmitting sensor data, forecasts, and alerts that transform slow processes into near-real-time feedback loops. Multi-system resilience requires structural rather than incremental change in how infrastructure is designed, managed, and connected to wider governance systems (Asadzadeh et al., 2022). When embedded in a systems framework that exposes interdependencies and aligns interventions with structural shifts, digital flows enable decision-making that links equity, resilience, and investment credibility. AI, IoT, and digital twin models can convert raw data into indicators of modal shift, building energy use, or air quality, enabling continuous learning. IPCC Assessment Report (AR)6 (2023) and the 2024 GRAA emphasize that coupling digitalization with communication, transparent data systems and institutional capacity converts compliance-driven reporting into opportunity-driven governance.

Trust and accountability determine value. Verification protocols, third-party audits, and open standards enable councils and investors to interpret results with confidence. Leading platforms tag provenance for each dataset and reveal calculation steps through open interfaces, applying the "inspectable model" principle advanced in transition-planning frameworks (Shalit et al., 2025). Differential data-sharing models that aggregate dashboards for public view and authenticate research access for detail protect privacy while sustaining transparency. Layered access prevents both surveillance overreach and data hoarding, issues cited in AR7 as sources of public resistance.

AI governance remains challenging. Algorithmic opacity and vendor dependence can shift power away from cities lacking technical expertise. Municipalities are adopting ethics frameworks that mandate explainability audits, fairness metrics, and retraining with representative data. Public–private data agreements now include "sunset clauses" transferring anonymized models back to cities, avoiding lock-in while preserving learning. Regulators are urged to reserve bandwidth and cloud resources for essential climate data rather than for commercial algorithms. Embedding public-interest obligations—open interfaces, audit rights, and non-exclusive licenses—into contracts balances innovation with accountability.

Knowledge gaps also persist across scales. Global emissions models often stop at metropolitan resolution, while fine-scale adaptation data, surface permeability, social vulnerability, and informal-settlement mapping remain local or analog. Open-data charters help bridge this gap by combining sensor and government data with citizen-generated information, legitimizing informal evidence (Conrad & Hilchey, 2011; Goodchild & Glennon, 2010). Machine-learning quality checks can validate crowdsourced inputs to official standards (Ponti & Craglia, 2020). Long-term consistency is essential: adaptable data schemas must absorb new technologies without rewriting governance rules.

Systems thinking clarifies why continuity matters. In Meadows's iceberg model, visible events rest on deeper patterns and structures (Meadows, 2008). Without longitudinal datasets, planners misdiagnose drivers such as induced demand or rebound energy use. When readiness metrics—capacity, permitting time, and citizen adoption—are tracked alongside emissions, cities can reallocate resources early. Evidence from 32 Scottish authorities shows that embedding such metrics in shared platforms links budgets to systemic milestones, accelerating blended-finance approvals (Bai et al., 2016; Shalit et al., 2025).

Looking beyond 2025, priorities are:

1. *Modular taxonomies* that integrate new technologies (edge AI, quantum optimization, hyperspectral imaging) without fragmenting standards.
2. *Inclusive digital strategies* combining in-person literacy programs and translation into local languages to close digital divides.
3. *Research on operations* and how to fund long-term data stewardship, quantify social co-benefits in machine-readable form, and federate community dashboards into national inventories without losing local nuance.

Digitally enabled climate governance will succeed only when systemic insight and social legitimacy progress together. Integrating open standards, participatory data, and ethics-by-design turns fragmented compliance into opportunity-driven transformation, building equitable and finance-ready pathways delivering climate-resilient development for all residents.

8.7 Conclusion and Recommendations

Digitalization now underpins how cities plan, finance, and implement climate action, but its success depends on inclusive governance and ethical integration. Tools gain value only when embedded in institutions that protect trust, accessibility, and long-term resilience.

The GRAA provides a framework through its eight dimensions, or "towers" -communication; decentralization; digitalization; efficiency; information integrity and transparency; risk; scale; and uncertainty—that help cities align data governance with systemic thinking. Insights from the Innovate4Cities conferences and the cases presented here reveal a growing demand for participatory, finance-ready digital tools to address ongoing deficits in standards and capacity, especially in smaller municipalities.

Data pluralism is vital. Combining digital and analog, formal and informal knowledge strengthens legitimacy and resilience. AI forecasts, satellite models, energy-use records, crowdsourced mobility data, oral histories, and community observations all form complementary evidence streams for decision-making. Embracing this diversity transforms data from a technical input into a democratic asset.

As the digital ecosystem expands, governance must evolve to safeguard the public interest. Cities require protocols for algorithmic accountability and measures that prevent monopolization of data infrastructure.

Capacity-building should move beyond digital literacy toward ethical reasoning, inclusive facilitation, and open-data stewardship. Leadership must acknowledge that every technical choice has social and political dimensions.

The path forward demands coordinated action across governments, financiers, and civil society. Table 8.1 consolidates practical recommendations drawn from the chapter. Each aligns actors with measurable progress indicators.

Together, these actions reposition data as a driver of transformation. When digitalization aligns with inclusive governance, systemic planning, and sustained capacity, it becomes a multiplier, accelerating climate action, making it fairer and more accountable.

Table 8.1 Recommendations from this Chapter

Recommendation	*Primary audience*	*Example KPI/Evidence of progress*
Adopt inclusive, technology-agnostic data-governance frameworks	City governments, national ministries	Number of climate indicators sourced from informal, analog, or citizen-generated data
Fund local capacity for gathering and interpreting non-digital data	Donor agencies, municipal finance departments	Number of participatory data initiatives funded annually per city
Embed technical- and data-literacy roles within local governments	City governments, HR departments	Number of full-time data-stewardship roles per 100,000 residents
Integrate co-produced knowledge into official decision-making processes	Local planners, policy designers	Presence of co-produced data in official climate adaptation or mitigation plans (Yes/No)
Develop ethics-based, public-interest digital infrastructure	National regulators, platform providers	Presence of smart-city tools with published ethics and data-privacy frameworks
Create differential data-sharing protocols that balance privacy and access	IT departments, civil-society watchdogs	Existence of tiered data-sharing policy with role-based access (Yes/No)
Prioritize modular, open-source digital tools	City IT departments, procurement offices	Presence of digital tools using open standards or open-source components
Ensure long-term investment in digital-infrastructure resilience	Ministries of ICT, donor agencies	Share of digital tools with multi-year upgrade and maintenance budgets (%)

(continued)

Table 8.1 (continued)

Recommendation	*Primary audience*	*Example KPI/Evidence of progress*
Make global climate data locally usable through downscaling and translation	Global dataset providers, research institutions	Presence of city plans using localized versions of global climate scenarios
Establish cross-departmental and multi-stakeholder data-sharing platforms	Local and regional governments	Number of departments with shared data access or integrated dashboards
Tie climate finance to transparent, public data dashboards	Finance ministries, investors, donor agencies	Value of climate funding tracked via publicly accessible performance dashboards
Support accessible data tools for small and under-resourced cities	Donors, national platforms, networks like ICLEI	Number of small cities receiving subsidized or free access to data platforms
Govern the use of AI and advanced analytics through public-interest safeguards	City governments, AI developers	Presence of AI audit trail, bias mitigation, and ethics policy (Yes/No)

Competing Interests The authors have no conflicts of interest to declare.

References

Ambole, A., Anditi, C., & Oni, T. (2025). Potential of citizen science to advance urban planetary health research in low and middle-income countries: A scoping review. *PLOS Global Public Health, 5*(5), Article e0003958. https://doi.org/10.1371/journal.pgph.0003958

Argyroudis, S. A., Mitoulis, S. A., Chatzi, E., Baker, J. W., Brilakis, I., Gkoumas, K., Vousdoukas, M., Hynes, W., Carluccio, S., Keou, O., Frangopol, D. M., & Linkov, I. (2022). Digital technologies can enhance climate resilience of critical infrastructure. *Climate Risk Management, 35*, Article 100387. https://doi.org/10.1016/j.crm.2021.100387

Asadzadeh, A., Khavarian-Garmsir, A. R., Sharifi, A., Salehi, P., & Kötter, T. (2022). Transformative resilience: An overview of its structure, evolution, and trends. *Sustainability, 14*(22), 15267. https://doi.org/10.3390/su142215267

Ataman, C., Herthogs, P., Tunçer, B., & Perrault, S. (2025). From insight to action: An integrated assessment framework for digital citizen participation in data-centric urban practices. *Cities, 156*, Article 105545. https://doi.org/10.1016/j.cities.2024.105545

Bai, X., Surveyer, A., Elmqvist, T., Gatzweiler, F. W., Güneralp, B., Parnell, S., Prieur-Richard, A., Shrivastava, P., Siri, J. G., Stafford-Smith, M., Toussaint, J., & Webb, R. (2016). Defining and advancing a systems approach for sustainable cities. *Current Opinion in Environmental Sustainability, 23*, 69–78.

Balogun, A.-L., Marks, D., Sharma, R., Shekhar, H., Balmes, C., Maheng, D., Arshad, A., & Salehi, P. (2020). Assessing the potentials of digitalization as a tool for climate change adaptation and sustainable development in urban centres. *Sustainable Cities and Society, 53*, Article 101888. https://doi.org/10.1016/j.scs.2019.101888

Bastidas, V., Oti-Sarpong, K., Nochta, T., Wan, L., Tang, J., & Schooling, J. (2024). Leadership of urban digital innovation for public value: A competency framework. *IET Smart Cities, 6*(3), 237–252. https://doi.org/10.1049/smc2.12063

Batty, M. (2018). Digital twins. *Environment and Planning B: Urban Analytics and City Science, 45*(5), 817–820. https://doi.org/10.1177/2399808318796416

Bortolini, R., Rodrigues, R., Alavi, H., Dalla Vecchia, L. F., & Forcada, N. (2022). Digital twins' applications for building energy efficiency: A review. *Energies, 15*(19), 7002. https://doi.org/10.3390/en15197002

Bozkurt, Y., Rossmann, A., Pervez, Z., & Ramzan, N. (2025). Development and evaluation of an urban data governance reference model based on design science research. *Government Information Quarterly, 42*(2), Article 102025. https://doi.org/10.1016/j.giq.2025.102025

Bulkeley, H., Broto, V. C., & Edwards, G. A. (2014). *An urban politics of climate change: Experimentation and the governing of socio-technical transitions*. Routledge.

Colona, F. (2023). Climate governance by numerical data: The kaleidoscopic political space of a decarbonization dashboard. *Geoforum, 144*, Article 103801. https://doi.org/10.1016/j.geoforum.2023.103801

Conrad, C. C., & Hilchey, K. G. (2011). A review of citizen science and community-based environmental monitoring: Issues and opportunities. *Environmental Monitoring and Assessment, 176*(1), 273–291. https://doi.org/10.1007/s10661-010-1582-5

de Vos, A., Preiser, R., & Masterson, V. (2021). Participatory data collection. In O. M. Schlüter & S. Beichler (Eds.), *The Routledge handbook of research methods for social-ecological systems* (pp. 95–106). Routledge. https://doi.org/10.4324/9781003021339-10

de Vries, A. (2023). The growing energy footprint of artificial intelligence. *Joule, 7*(10), 2191–2194. https://doi.org/10.1016/j.joule.2023.09.004

Gebremedhin, E. T., Basco-Carrera, L., Jonoski, A., Iliffe, M., & Winsemius, H. (2020). Crowdsourcing and interactive modelling for urban flood management. *Journal of Flood Risk Management, 13*(2), Article e12602. https://doi.org/10.1111/jfr3.12602

Global Covenant of Mayors for Climate and Energy and UN-Habitat. (2022). Discussions and findings from the 2021 Innovate4Cities conference: New insights for the 2018. In B. Walsh & J. Greenwalt (Eds.), *Global research and action agenda on cities and climate change science*. https://doi.org/10.26188/62795d16f3bab

Goodchild, M. F. (2007). Citizens as sensors: The world of volunteered geography. *GeoJournal, 69*, 211–221. https://doi.org/10.1007/s10708-007-9111-y

Goodchild, M. F., & Glennon, J. A. (2010). Crowdsourcing geographic information for disaster response: A research frontier. *International Journal of Digital Earth, 3*(3), 231–241. https://doi.org/10.1080/17538941003759255

Haasnoot, M., Kwakkel, J. H., Walker, W. E., & Ter Maat, J. (2013). Dynamic adaptive policy pathways: A method for crafting robust decisions for a deeply uncertain world. *Global Environmental Change, 23*(2), 485–498.

Haklay, M. (2012). Citizen science and volunteered geographic information: Overview and typology of participation. In *Crowdsourcing geographic knowledge: Volunteered geographic information (VGI) in theory and practice* (pp. 105–122). Springer Science and Business Media.

Helmi, A., Bastidas, V., Oti-Sarpong, K., & Schooling, J. (2024). Sustainable urban digital innovation: A socio-technical competency-based approach to evaluation. *Sustainable Cities and Society, 117*, Article 105946. https://doi.org/10.1016/j.scs.2024.105946

Hsu, A., Höhne, N., Kuramochi, T., Roelfsema, M., Weinfurter, A., Xie, Y., Lütkehermöller, K., Chan, S., Corfee-Morlot, J., Drost, P., & Faria, P. (2019). A research roadmap for quantifying non-state and subnational climate mitigation action. *Nature Climate Change, 9*(1), 11–17.

Huang-Lachmann, J.-T. (2019). Systematic review of smart cities and climate change adaptation. *Sustainability Accounting, Management and Policy Journal, 10*(4), 745–772. https://doi.org/10.1108/SAMPJ-03-2018-0052

Hughes, S., Giest, S., & Tozer, L. (2020). Accountability and data-driven urban climate governance. *Nature Climate Change, 10*, 1085–1090. https://doi.org/10.1038/s41558-020-00953-z

Hunter, N., Walsh, B., Ramos, G. C. D., Greenwalt, J., Roberts, D., Oke, C., Nielsen, D., & Slotow, R. (2023). Cities and climate change science after Edmonton: Knowledge, people, and organisations' influence and outreach. *Current Research in Environmental Sustainability, 6*, Article 100231.

ICLEI, C40, & World Resources Institute. (2014). *Global protocol for community-scale greenhouse gas emission inventories (GPC)*. http://www.ghgprotocol.org/city-accountinghttps://ghgprotocol.org/sites/default/files/ghgp/standards/GHGP_GPC_0.pdf

IPCC. (2023). Sections. In H. Lee & J. Romero (Eds.), *Climate Change 2023: Synthesis report. Contribution of working groups I, II and III to the sixth assessment report of the intergovernmental panel on Climate Change* [Core Writing Team] (pp. 35–115). IPCC. https://doi.org/10.59327/IPCC/AR6-9789291691647

Irvin, A., Oke, C., Khan, S., Jance IV, B., Ranalder, L., Barth, B., Deacon, A., Abusamra, R., Addison, A., Adebisi, J., Adsett, C., Ailio, E., Al Jadaa, D., Aslan, S., Assalini, S., Assarkhaniki, Z., Astolfo, J. P., Badino, M., Bai, …, & Williams, A. (2024). *Next-generation city climate action: Findings from the 2024 Innovate4Cities conference and update to the global research and action agenda on cities and climate change science (GRAA)*. Global Covenant of Mayors for Climate and Energy, UN-Habitat & University of Melbourne. https://doi.org/10.26188/27147612

Jackson, E., Jones, A., & Salehi, P. (2022). Advancing community resilience through smart approaches: A resource for Canadian communities. In *Resilient smart cities: Theoretical and empirical insights* (pp. 443–458). Springer International Publishing.

Law, K. H., & Zhang, D. (2018). Digital twin and big data towards smart manufacturing and Industry 4.0: 360 degree comparison. *IEEE Access, 6*, 3585–3593. https://doi.org/10.1109/ACCESS.2018.2793265

Lember, V., Brandsen, T., & Tõnurist, P. (2019). The potential impacts of digital technologies on co-production and co-creation. *Public Management Review, 21*(11), 1665–1686. https://doi.org/10.1080/14719037.2019.1619807

Lonergan, K. E., Köll, J. F., & Sansavini, G. (2025). Model-based Climate Action Plans for ambitious local emissions reduction: A project-focused approach. *Urban Climate, 61*, Article 102452. https://doi.org/10.1016/j.uclim.2025.102452

Meadows, D. H. (2008). *Thinking in systems: A primer*. Chelsea Green.

Mendonça, F. M., Souza, J. F., & Soares, A. L. (2024). Analysis dimensions for understanding digital twins: A discussion in the Brazilian scenario. *Revista Gestão & Tecnologia. Journal of Management and Technology, 24*(1), 91–111.

Monasterolo, I., & Battiston, S. (2022). Sustainable investing and climate transition risk: A portfolio rebalancing approach. *Journal of Portfolio Management, 48*(1). https://doi.org/10.3905/jpm.2022.1.394

Mukim, M., & Roberts, M. (Eds). (2023). *Thriving: Making cities green, resilient, and inclusive in a changing climate*. World Bank. http://hdl.handle.net/10986/38295

Nguyen, T. A., Garcia, L., & Dinh, H. (2025). A wearable sensor node for measuring air quality through citizen participation. *Sensors, 25*(4), 2211.

Nick, F. C., Sänger, N., van der Heijden, S., & Sandholz, S. (2023). Collaboration is key: Exploring the 2021 flood response for critical infrastructures in Germany. *International Journal of Disaster Risk Reduction, 87*, Article 103710. https://doi.org/10.1016/j.ijdrr.2023.103710

OECD/The World Bank/UN Environment. (2018). Financing climate futures: Rethinking infrastructure. *OECD Publishing*. https://doi.org/10.1787/9789264308114-en

Ponti, M., & Craglia, M. (2020). *Citizen-generated data for public policy*. European Commission, Ispra. JRC120231.

Salehi, P., & Sharifi, A. (2022). Toward integrating resilience thinking in smart city planning and development. In *Resilient smart cities: Theoretical and empirical insights* (pp. 459–466). Springer International Publishing.

Sanchez, T. W., Brenman, M., & Ye, X. (2024). The ethical concerns of artificial intelligence in urban planning. *Journal of the American Planning Association, 91*(2), 294–307. https://doi.org/10.1080/01944363.2024.2355305

Scottish Climate Intelligence Service. (n.d.). *Multi-level climate governance in Scotland*. https://www.climateintelligenceservice.scot/__data/assets/pdf_file/0023/62555/SCIS_Multi-Level-Governance_Case-Study.pdf

Shalit, T., Dixon, M. & Bergöö, M. (2024). *Standardising climate mitigation: The transition element framework: An open-source initiative to structure and codify IPCC knowledge for implementation of climate action*. ClimateView AB.

Shalit, T., Dixon, M., Corbett, D., Salehi, P., Schultz, L., Manner, M., Andrén, C., Wharmby, C., & Boertien, R. (2025). *Designing for delivery: How cities can structure fundable climate transitions* (Version 1.0). ClimateView AB. https://www.climateview.global/en/whitepaper/designing-for-delivery

Shen, Y., Yang, Z., & Zhang, X. (2023). Impact of digital technology on carbon emissions: Evidence from Chinese cities. *Frontiers in Ecology and Evolution, 11*, Article 1166376. https://doi.org/10.3389/fevo.2023.1166376

Soden, R., Wagenaar, D., & Tijssen, A. (2021). *Responsible AI for disaster risk management: Working group summary*. Global Facility for Disaster Reduction and Recovery (GFDRR), Deltares, & University of Toronto. https://opendri.org/wp-content/uploads/2021/06/ResponsibleAI4DRM.pdf

Taylor, A., Jack, C., McClure, A., Bharwani, S., Ilunga, R., & Kavonic, J. (2021). Understanding and supporting climate-sensitive decision processes in southern African cities. *Current Opinion in Environmental Sustainability, 51*, 77–84. https://doi.org/10.1016/j.cosust.2021.03.006

United Nations Development Programme. (2021). *Tuvalu coastal adaptation project*. https://www.undp.org/pacific/projects/tuvalu-coastal-adaptation-project

United Nations Framework Convention on Climate Change. (2021). *Human needs-based climate innovation framework*. Global Innovation Hub.

van Berlo, L., Rutherford, P., & Al-Hussein, M. (2020). Towards a semantic construction digital twin: Directions for future research. *Automation in Construction, 114*, Article 103179. https://doi.org/10.1016/j.autcon.2020.103179

Verhulst, S. G. (2024). The need for climate-data stewardship: Ten tensions and reflections regarding climate-data governance. *Data & Policy, 6*, Article e52. https://doi.org/10.1017/dap.2024.52

West, S., van Kerkhoff, L., & Wagenaar, H. (2019). Beyond "linking knowledge and action": Towards a practice-based approach to transdisciplinary sustainability interventions. *Policy Studies, 40*(5), 534–555. https://doi.org/10.1080/01442872.2019.1618810

Wilby, R. L., Dawson, C. W., Murphy, C., O'Connor, P., & Hawkins, E. (2014). The Statistical DownScaling Model – Decision Centric (SDSM-DC): Conceptual basis and applications. *Climate Research, 61*, 259–276. https://doi.org/10.3354/cr01254

World Health Organization (WHO) Regional Office for Africa. (2024). *Small Island developing states pooled procurement initiative in the WHO African Region: From declaration to actions and results*. https://www.afro.who.int/sites/default/files/2024-09/SIDS20Pooled20Procurement-23sept2420Vers202.pdf. Accessed October 30, 2025.

Yan, L., & Cheng, D. (2024). Opportunities and challenges for urban climate governance. *Nature Climate Change, 14*(11), 1112–1115. https://doi.org/10.1038/s41558-024-02173-1

Yang, C., Gu, M., & Albitar, K. (2024). Government in the digital age: Exploring the impact of digital transformation on governmental efficiency. *Technological Forecasting and Social Change, 208*, Article 123722. https://doi.org/10.1016/j.techfore.2024.123722

Zewe. (2025). *Explained: Generative AI's environmental impact*. MIT News.

PART III

Systems Approaches to City Research and Innovation

The imperative to rethink urban systems stems not only from climate and environmental concerns but also from the convergence of multiple, interlocking crises that are reshaping our cities at every level. The accelerating climate emergency, widening socioeconomic divides, and increasingly strained public-health infrastructures all expose the limits of fragmented urban planning and governance approaches rooted in sectoral logic and linear thinking. In response, a systemic perspective—one that acknowledges the city as a complex, adaptive system where infrastructures, institutions, and everyday practices are deeply interconnected—becomes not just desirable but necessary.

Infrastructures, in turn, must be reimagined beyond their traditional technical definition. This means viewing them not merely as physical systems—pipes, roads, grids—that move resources and expel waste, but as interconnected socioeconomic, ecological, and civic systems that sustain collective life. Libraries, parks, care centers, and digital commons, for instance, are equally vital components of urban infrastructure because they enable learning, care, cultural exchange, and civic participation. These are not "soft" add-ons, but essential foundations of resilience, equity, justice, and democratic social cohesion.

Accelerating urban transformation requires moving beyond siloed governance structures and adopting a systemic, inclusive approach that embraces ongoing public participation, acknowledges complexity, accounts for trade-offs and co-benefits, and embraces a more ecological and human-centered perspective.

The challenge involves building more efficient, smart, and circular cities that are also livable, just, and inclusive. This entails advancing spatial and temporal justice to support a more balanced urban economic and social order, while enhancing sustainability. It also requires delineating transformational urban pathways that anticipate potential climateovershoot scenarios, while advancing the vision of cities that are regenerative and resilient. Regenerative cities enhance and restore the systems they depend on, and resilient cities can adapt to future shocks and stresses—including compound risks and cascading impacts—yet remain grounded in the everyday realities, needs, and aspirations of their inhabitants.

The GRAA, which aims to guide science and policy toward better adaptation and mitigation of climate change in cities, has been framed since its conception as a systemic approach from which key action priorities, cross-cutting issues, solutions, and knowledge gaps are identified. The 2024 GRAA describes 12 systemic elements for empowering urban transformation aimed at climate-resilient development. For analytical purposes, those elements have been organized here into four transformation realms: land and ecosystems, urban physical infrastructure and public services, urban resource flows, and socioeconomic and governance structures.

The following three chapters on systems approaches to city research and innovationdiscuss the GRAA systemic elements (Table 1). Chapter 9 addresses the realm of land and ecosystems, which primarily refers to the biophysical context, examined through the lenses of biodiversity, culture, and geography. Chapter 10 explores the realms of the built environment and urban services, as well as urban resource inflows and outflows, analyzed from the perspective of infrastructure, resource efficiency, public services, and health. Finally, Chapter 11 focuses on the realm of socioeconomic and governance structures, providing a review of finance, governance, and multi-level partnerships.

Table 1 Realms and systemic elements for urban sustainable and climate-resilient

Transformation realm	*Related GRAA systemic elements*	*Main chapter addressing GRAA elements*
Land and ecosystems (the biophysical context)	(1) Geography (2) Biodiversity (3) Health (environmental)	Chapter 9 Chapter 10from the perspective of land uses and urban planning
Urban infrastructure and public services (the built-environment context)	(3) Infrastructure (including buildings and green infrastructure) (4) Mobility (5) Water (6) Energy (7) Food (8) Waste (9) Health (human and environmental)	Chapter 10 Chapter 9 addressing nature base solutions in relation to health
Urban resource inflows and outflows		
Socioeconomic and governance structures (including public outreach and education)	(10) Culture (Community, Heritage, Art and History) (11) Governance and multi-level partnerships (12) Finance	Chapter 11 Chapter 10 only as a brief consideration of their relevance to the social context (or social infrastructure)

Source Authors

CHAPTER 9

Systems Approaches to City Research and Innovation: Biodiversity, Culture, and Geography

Niki Frantzeskaki, Judy Bush, Perrine Hamel, Melissa Pineda-Pinto, Timon McPhearson, Dave Kendal, Marcus J. Collier, and Johanna Granados Alcala

9.1 Introduction

Cities are complex social—ecological—technological systems (SETS). A SETS framing highlights the interactions and interrelations in cities between people, biodiversity and ecosystems, and the built form and its

N. Frantzeskaki (✉)
Utrecht University, Utrecht, The Netherlands
e-mail: n.frantzeskaki@uu.nl

J. Bush
Faculty of Architecture Building and Planning, University of Melbourne, Melbourne, Australia
e-mail: judy.bush@unimelb.edu.au

C. Oke et al. (eds.), *A Global Research and Action Agenda on Cities and Climate Change Science*,
https://doi.org/10.1007/978-981-95-6517-7_9

tools and urban infrastructures (McPhearson et al., 2022). This systems approach emphasizes understanding the dynamic interactions and feedback mechanisms among these domains, recognizing that the whole is greater than the sum of its parts, and that changes in one sphere inevitably lead to emergent properties and cascading effects across the entire system (Chester et al., 2023). Its primary purpose is to move beyond sectoral analyses to inform integrated decision-making, ensuring that interventions promote resilience and sustainability without causing unintended negative consequences to the system's overall function and service provision. However, cities are often not evaluated, planned, or governed as

---

P. Hamel
Asian School of the Environment, Nanyang Technological University, Singapore, Singapore
e-mail: perrine.hamel@ntu.edu.sg

M. Pineda-Pinto
Melbourne Centre for Cities, Faculty of Architecture Building and Planning, The University of Melbourne, Melbourne, Australia
e-mail: melissa.pinedapinto@unimelb.edu.au

T. McPhearson
Urban Systems Lab, New York University, New York, NY, USA

Cary Institute of Ecosystem Studies, Milllbrook, NY, USA

Beijer Institute of Ecological Economics at the Swedish Royal Academy of Sciences, Stockholm, Sweden

T. McPhearson
e-mail: timon.mcphearson@nyu.edu

D. Kendal
Future in Nature, Melbourne, Australia
e-mail: dave@futurenature.au

M. J. Collier
Trinity College Dublin, University College Dublin, Dublin, Ireland
e-mail: marcus.collier@tcd.ie

J. Granados Alcala
Global Covenant of Mayors, Brussels, Belgium
e-mail: jgranadosalcala@gcomprojectsupport.org

such. This results in cities being imagined and conceived in less interconnected ways that overlook or ignore complexity, with fragmented governance systems and planning processes that produce undesired trade-offs, and that fail to consider cross-sector and cross-scale impacts. These underlie the knowledge gaps presented in the GRAA 2024 report. These knowledge gaps span local contextualization and institutional fitness of climate and biodiversity plans, issues of regional connectedness, and design, testing, and evidence-based planning for tailored urban solutions.

At the same time, to address these knowledge gaps, a growing number of urbanists conceptualize cities and urbanized regions as tightly intertwined and complex SETS (McPhearson et al., 2021, 2022, 2023; Chester et al., 2023; McPhearson et al., 2022). The SETS conceptual framework applied to urban systems suggests increased attention to the dynamic interactions between the social-economic, ecological, and technological-infrastructural domains of cities that drive patterns and processes affecting daily life and shape urban futures. A systems approach, whether applied through the SETS framework or others (Zhou et al., 2021), is critical to reveal new opportunities for innovation at the intersection of systems that are normally governed, planned, and managed in isolation.

Taking a systems approach to understanding urban patterns and urban change, as well as to designing urban plans and solutions, requires a focus on complexity. Complexity exists in both the diversity of urban system elements (land uses, governance, communities, biodiversity, health, education, and transport) and in their interactions (or nexus between elements): the interrelations of people, infrastructure, nature/ecosystems, and time in cities. Recent work on systemic approaches (including the IPBES Nexus Assessment 2025, discussed in O'Brien et al., 2025) puts an emphasis on understanding the interrelations between systems of provision and services. A nexus understanding and conceptualization, particularly within the SETS context, that focuses on how these individual system elements interact, interrelate, and affect each other may progress urban system thinking toward innovative conceptualizations and solutions. A nexus approach can potentially inform systemic understandings of city functions and urban change, as well as drive policy development and practice that recognizes these interrelations and interactions, which are inherent in cities.

We propose nexus thinking as a way to innovate across scales (addressing regional connectivity) and across sectors (to tailor solutions that are cross-sectoral and inclusive) to address the polycrisis: the interlinked, compounding threats posed by biodiversity loss, climate change, inequity and injustice. Innovation refers to policy, planning, and governance actions that are novel in their context and push forward for a systemic and cross-sectoral (hence nexus) thinking and doing, for more climate-resilient and nature-positive urban futures. Drawing from the knowledge gaps identified in the 2024 GRAA (Irvin et al., 2024), we summarize key gaps and recent advances in research and practice to address these gaps, with a focus on three themes: biodiversity, culture, and geography (other chapters in this book explore complementary themes).

9.2 Challenges and Opportunities

For cities innovating through greater engagement with nature, we need to consider the following nexus challenges and opportunities:

9.2.1 Biodiversity

The biodiversity knowledge gaps identified in the latest Innovate4Cities initiative report (Irvin et al., 2024) relate to the key challenge of policy "dis-integration": how to integrate biodiversity conservation and restoration into urban planning and development. Ecological degradation is driven by multiple human-induced impacts. The compounded factors of climate change, invasive species, habitat fragmentation, and pollution are exacerbating rapid biodiversity loss, with cities as fundamental drivers of these patterns through urban development, resource consumption, and greenhouse gas emissions (IPBES, 2019; Parmesan et al., 2022; Pörtner et al., 2021). Urban areas, including the species, community relationships, and larger social and infrastructure contexts in which urban nature exists, are facing multiple threats from human-mediated disturbances (McPhearson et al., 2023). This ecological degradation is exacerbated by environmental stressors and the changes in land-use patterns that come with urban development.

Urban nature provides a wide range of ecological, social, and economic benefits. Ecologically, green and blue infrastructure improves air and water quality, mitigates urban heat-island effects, and supports pollination and biodiversity corridors. Small patches of vegetation in cities create a network that serves as a critical habitat for birds, insects, and small mammals (Elmqvist et al., 2013). Socially, urban green spaces contribute to public health by encouraging physical activity and reducing stress (Tzoulas et al., 2007). Well-planned and located nature-based solutions and green spaces that mitigate urban heat can contribute to cities' capacity to support vulnerable communities who have less access to cooler, high-quality housing (Bush, 2025). They also foster attachment and promote social interaction. Urban biodiversity reflects and reinforces First Peoples' holistic approaches to sustainability, custodianship, and resource management (Cumpston, 2020). Urban nature has cultural and spiritual dimensions, which contribute to building and maintaining a local identity. Economically, proximity to green areas increases property values and reduces energy costs by providing shade and cooling. Green urban areas host agricultural projects that are vital in some communities for food security, community engagement, and empowerment.

Irvin et al. (2024) highlight the importance of vibrant and active cities, advocating for governance models that integrate multiple decision-making levels and actors, where communities are active contributors, ensuring that community values are represented in urban planning efforts. Urban-biodiversity strategies must thus be tailored to local environmental, social, and economic contexts. Place-based responses should incorporate local knowledge, past and present culture, art, and heritage to address the community's needs.

Aligning and localizing multi-level governance are key to addressing policy dis-integration. First, there is a misalignment of regulatory governance at local, national, and global levels for biodiversity, making urban goals for biodiversity neither systematically prioritized nor addressed across different urban sectoral policies and plans. This cross-scale misalignment is a multi-level governance challenge: cities are sites in which multi-level governance must work to align local and national priorities such as climate change (Bergman et al., 2025) and biodiversity. Second, there is a need for coordination of actors at different levels of governance for the integration and localization of biodiversity in policies and plans, as well as recognizing and addressing in policy plans the connected needs between biodiversity, climate change, and resilience (McPhearson et al., 2025).

To effectively address these challenges, there is an urgent need for strategies that integrate biodiversity conservation into urban planning and development, and that address the critical nexus between biodiversity and culture. In multi-level governance contexts, this requires and is enabled by bi-directional flows of information from national to local and also from cities to national and global levels. Such strategies should focus on protecting vegetation, enhancing green spaces, and implementing policies that strike a balance between development and environmental preservation, and that bring diverse knowledge and cultural values directly into biodiversity and ecological planning and governance. Traditional approaches to governance have tended to ignore many other ways of learning, knowing, and doing, and struggle to connect to local and regional cultural practices (Gazing Wolf et al., 2024; Orlove et al., 2023). Indigenous, local, and Traditional knowledge systems contain vast amounts of critical information that needs to be seen, heard, and used to work with other tools and strategies from Western science (Bartlett et al., 2012; Nelson & Shilling, 2018). Deep, local, and cultural relationships with urban ecosystems developed over generations and embedded within Indigenous knowledge systems need to be more actively integrated into approaches to urban development to build cities that enhance, rather than threaten, biodiversity (Cumpston et al., 2022).

9.2.2 Culture

The culture knowledge gaps identified in the Innovate4Cities initiative report (Irvin et al., 2024) highlight a key challenge in integrating cultural and historical considerations into urban development and climate action. This includes the need for better coordination in urban planning to equitably preserve cultural heritage, effectively incorporating community perspectives into heritage management, and understanding the impact of urbanization on cultural sites. Furthermore, there is a significant gap in fully integrating culture, art, heritage, and history into participatory planning processes for climate change solutions, which, if integrated, will ensure the inclusion of marginalized communities and recognize past injustices, and apply cultural insights for effective, place-based climate responses, including the incorporation of oral histories (Crowley et al., 2022).

Climate change is raising awareness of the dynamic and diverse nature of culture and heritage. Rather than automatic preservation of historical norms and baselines, processes that recognize that heritage and culture must be allowed to change are being adopted. A key mechanism used to put this understanding into practice is through community engagement (Waterton, 2015). Rather than assigning cultural value or historic importance a priori, or assuming power through expertise and scientific understanding, community engagement allows multiple ways of understanding cultural value and prioritizing heritage to be revealed. This can aid the legitimacy of cultural and heritage processes. Key advances include theorizing community in ways that allow complex and fluid community identities, and recognizing the importance of benefits to participants in community-engagement processes. Genuine community engagement with diverse communities promises more human-centered outcomes, rather than technocratic outcomes (Gulsrud et al., 2018). These processes are being used to acknowledge and weave Traditional Knowledge systems and worldviews into decision-making and to support self-determination in settler-colonial societies such as Australia, New Zealand, and Canada (Bartlett et al., 2012; Waterton, 2015).

Progress toward addressing this gap from a justice perspective has seen some promising shifts. For example, the climate and multispecies justice frameworks demonstrate a crucial pivot toward integrating justice considerations into climate change discourse and action, moving beyond purely technical solutions to embrace socio-cultural, ethical, and historical dimensions. Climate justice, in particular when taking an intersectional perspective, empowers us to recognize and confront the interconnected social and environmental injustices that create urban vulnerabilities (Amorim-Maia et al., 2022). This, in turn, fosters the development of more practical and integrated strategies for equitable urban adaptation and transformation (Mikaelsson & Lager, 2024). Progress in the field of multispecies justice—which extends justice beyond human concerns to encompass non-human entities and entire ecological systems—involves the conceptualization and emergence of practices that foster respectful relationships with all life on Earth, drawing significantly from Indigenous philosophies and Local Knowledge, challenging anthropocentric approaches to governance and development (Celermajer et al., 2021). Efforts are being made to transform political, social, cultural, legal, and economic systems to be more responsive to the needs and rights of human and more-than-human beings. For example, the Nature Futures Framework is gaining traction as a way to explore and clarify the multiple dimensions of nature-related values—including those connected to more-than-human beings—in urban contexts (Mansur et al., 2022). It provides a useful basis for dialogue on how people perceive and engage with nature and culture in cities.

To integrate cultural considerations into urban development and climate action effectively, breaking down isolation and horizontal-vertical jurisdictional barriers and challenging existing power structures requires a multi-faceted approach involving governance innovations, collaborative mechanisms, and a re-evaluation of how culture is valued and integrated. Ways to do this include mainstreaming the process of embedding climate-resilience considerations across different policy domains and governance levels (Adams et al., 2023). Such systemic integration could, in principle, create avenues for considering diverse cultural values and practices if the mainstreaming process itself is designed to be culturally sensitive and inclusive.

9.2.3 Geography

The geography knowledge gaps identified in the Innovate4Cities initiative report (Irvin et al., 2024) relate to our understanding of local geographical contexts for developing climate-responsive and nature-positive policies. Geography, the biophysical context, and the relationships between people, their environments, and their more-than-human neighbors are a foundational element in understanding the unique set of challenges and opportunities that each city faces (Castree et al., 2025). Geography, frequently referred to as local context, distinguishes one city from another (Irvin et al., 2024). Each city's geography has defined the city's form and development: circling mountains constrain urban expansion; rivers bring movement, transport, and flooding; climate, rainfall patterns, and landforms combine with evolving social dynamics and characteristics to shape how cities develop and function. While there is increasing research seeking to analyze and understand the "science of cities" (McPhearson et al., 2016) and develop informed approaches to urban sustainability and climate change (Frantzeskaki et al., 2021), key challenges remain in "localizing" this knowledge (Irvin et al., 2024): while facing similar challenges, cities are unique and responses must be adapted to local social-ecological-technological contexts. Although knowledge can be exchanged, it must always be contextualized. Further, there are substantial gaps in the global coverage of research findings, with less research focused on Global South cities (Chu et al., 2015) and smaller, regional urban settlements (Kendal et al., 2020).

There is growing research that emphasizes geography and local context in the elaboration of barriers and facilitators to address the polycrisis. An example is research for sustainability and climate change focused on the relationship between urban form and carbon emissions (Hurlimann et al., 2025), which points to the contributions achievable through transit-oriented design and active transport. Integrating nature-based solutions into urban areas can improve sustainable urban form and function: tree-lined streets encourage active transport and heat mitigation (Tang & Bush, 2025), sustainable drainage infrastructures contribute to stormwater management (Bush et al., 2023), and nature-based solutions and ecosystem-based adaptation can contribute to reducing the risk from disasters, including flooding and drought (Frantzeskaki & McPhearson, 2022), although there continue to be knowledge gaps in how local context influences implementation and ongoing maintenance.

While cities are sharing information on planning and implementing nature-based solutions via platforms such as CitiesWithNature (https://citieswithnature.org/), and through city-based processes such as the Berlin Urban Nature Pact (https://citieswithnature.org/berlin-urban-nature-pact/), there are continuing research efforts, yet also knowledge gaps, related to developing business cases and funding mechanisms for diverse local contexts (Collier et al., 2023). Urban intermediaries, including city-networks, using tools such as digital platforms, enable knowledge-sharing and open training that supports cities to learn from and adapt actions tested in other cities facing similar challenges.

Localizing urban research and ensuring the inclusion of diverse voices from biophysical and social sciences (Castree et al., 2014), First Peoples and local communities (Cumpston et al., 2022), and diverse urban contexts (West et al., 2024) is essential. Localizing urban knowledge and urban research and understanding how the surrounding regional contexts influence cities are pivotal to responding to urban complexity and to better understanding the urban nexuses for co-designing climate solutions. Co-design is an integrative participatory decision-making process that seeks to ensure inclusive engagement and integration of diverse views (O'Donnell et al., 2025). Examples include foregrounding Traditional Owners' aspirations by reintegrating fire into urban landscapes for ecosystem management and reconnecting wetlands to floodplains and seasonal flows (Cumpston et al., 2022). Further, local responses to climate change need to account for climate impacts in more just and inclusive responses that more effectively take account of capacity disparities, climate injustice/justice, and vulnerabilities. City planners, practitioners, and decision-makers need locally relevant research and insights, necessitating the collection and analysis of appropriate and meaningful indicators and data sets that span social and ecological factors (for example, Fanning & Raworth, 2025). There are opportunities for community science to contribute to this work, and in doing so to provide opportunities for broader participation and engagement (Tengö et al., 2021). Furthermore, initiatives such as "extreme community science" can help vulnerable communities achieve effective outcomes by improving data accessibility, fostering knowledge co-production, and building community capacity (English et al., 2018).

9.3 What Do We Need to Innovate With Nature in Cities?

In response to these challenges and to the identified knowledge and practice gaps, we take a systems approach through a nexus conceptualization and propose four ways forward. First, the adoption of integrated frameworks is critical so as to identify leverage points in the nexus of biodiversity—culture—geography/locality to move from problem identification to scaling innovative solutions. We propose to employ integrated frameworks to elucidate and map the complexity of the nexus in cities. Importantly, employing integrated frameworks is not intended to remove complexity but rather to provide tools to help navigate this complexity. Frameworks such as SETS, adaptive pathways, or the Nature Futures Framework (Pereira et al., 2020) show how to systematically map existing innovations, identify progressive pathways for action, and serve as organizing frameworks for transdisciplinary co-production of urban innovative solutions. Integrative frameworks can also aid urban researchers, planners, and practitioners to understand and embrace complexities as well as to identify synergies and trade-offs across the nexus to co-develop more-effective urban plans for innovating with nature in cities.

Second, innovative approaches to cities with nature, in the nexus of biodiversity—culture—geography, requires innovation in the concepts used to guide urban policy and planning. We thus propose that we need to progress narrative shifts toward more plural and decolonial understandings of innovations and the process (of what constitutes and what entails) innovating in cities with nature. Such plurality of understandings will allow for respect and acknowledgment of Traditional Owners' aspirations and custodianship of lands and waters. It will also open spaces for local communities (including vulnerable, marginalized groups from an intersectional justice perspective), and set up a platform for dialogue between knowledge holders. As language can connect and divide in urban visions, plurality, and inclusivity in terms may be a way to also pluralize innovations with nature in cities.

Third, we propose to integrate biodiversity, cultural, and geographical considerations into urban development, planning and governance for dealing with the polycrisis. By integrating "culture" and "nature" (human and more-than-human) in policy and planning for the polycrisis, we can both acknowledge cities as complex SETS, as well as explicitly and intentionally consider the policy interactions across policy domains, and the governance interactions across governance levels and between governance and policy actors. Furthermore, while colonial processes of urbanization have contributed to dispossession and habitat destruction, transformative approaches to planning and governance that embed cultural considerations and Local and Traditional Knowledges may more effectively challenge business-as-usual power imbalances and economic processes to progress action for the polycrisis (Frantzeskaki et al., 2024).

Fourth, instead of only using monitoring and evaluation to learn from what has been done, we propose to emphasize opportunities for applying lessons to develop innovative planning and development, taking an adaptive, forward-looking approach. In this way, learning across diverse urban contexts will serve two innovative pathways: to localize urban research and practice (policy and planning) and to avoid imagination lock-ins (associated with business-as-usual approaches). To support local monitoring and evaluation efforts, as well as connect these to global reporting processes, careful selection of indicators is required. It can be challenging to identify appropriate indicator sets that can effectively represent the complexity of ecosystem structure and function, within the broader urban fabric, and allow time and place comparisons. For such learning-forward practices to take off, strengthening dialogues and collaborations between researchers, practitioners, and communities is foundational. Collaborative learning processes that integrate citizen-science efforts with structured data collection, as well as diverse knowledge systems, can contribute to the fundamental processes of recognition, participation, co-learning, and co-management.

Acknowledgements The authors acknowledge Chiara Morfeo, of C40 Cities, for her contributions to the development of this chapter.

Competing Interests
The authors have no conflicts of interest to declare.

References

Adams, C., Frantzeskaki, N., & Moglia, M. (2023). Mainstreaming nature-based solutions in cities: A systematic literature review and a proposal for facilitating urban transitions. *Land Use Policy, 130*, Article 106661. https://doi.org/10.1016/j.landusepol.2023.106661

Amorim-Maia, A. T., Anguelovski, I., Chu, E., & Connolly, J. (2022). Intersectional climate justice: A conceptual pathway for bridging adaptation planning, transformative action, and social equity. *Urban Climate*, *41*, 101053. https://doi.org/10.1016/j.uclim.2021.101053

Bartlett, C., Marshall, M., & Marshall, A. (2012). Two-eyed seeing and other lessons learned within a co-learning journey of bringing together indigenous and mainstream knowledges and ways of knowing. *Journal of Environmental Studies and Sciences, 2*(4), 331–340. https://doi.org/10.1007/s13412-012-0086-8

Bergman, J., Buylova, A., Jacksson, J., Nasiritousi, N., Sanderink, L., & Tobin, P. (2025). The multilevel governance of polarised climate politics: How a pioneer city navigates a dynamic political context. *Environmental Politics*. https://doi.org/10.1080/09644016.2025.2533605

Bush, J. (2025). Planning nature-based solutions for cooling Australian cities. In P. B. Cobbinah & M. Addaney (Eds.), *Handbook on planning and climate change adaptation* (pp. 229–246). Edward Elgar Publishing. https://doi.org/10.4337/9781803926926

Bush, J., Frantzeskaki, N., Ossola, A., & Pineda-Pinto, M. (2023). Priorities for mainstreaming urban nature-based solutions in Australian cities. *Nature-Based Solutions, 100065*. https://doi.org/10.1016/j.nbsj.2023.100065

Castree, N., Adams, W. M., Barry, J., Brockington, D., Büscher, B., Corbera, E., Demeritt, D., Duffy, R., Felt, U., Neves, K., Newell, P., Pellizzoni, L., Rigby, K., Robbins, P., Robin, L., Rose, D. B., Ross, A., Schlosberg, D., Sörlin, S., & Wynne, B. (2014). Changing the intellectual climate. *Nature Climate Change, 4*(9), 763–768. https://doi.org/10.1038/nclimate2339

Castree, N., Barnes, T., & Salmond, J. (Eds.). (2025). *Making geography matter: The past and present of a changing discipline*. Routledge.

Celermajer, D., Schlosberg, D., Rickards, L., Stewart-Harawira, M., Thaler, M., Tschakert, P., Verlie, B., & Winter, C. (2021). Multispecies justice: Theories, challenges, and a research agenda for environmental politics. *Environmental Politics, 30*(1–2), 119–140. https://doi.org/10.1080/09644016.2020.1827608

Chester, M. V., Miller, T. R., Muñoz-Erickson, T. A., Helmrich, A. M., Iwaniec, D. M., McPhearson, T., Cook, E. M., Grimm, N. B., & Markolf, S. A. (2023). Sensemaking for entangled urban social, ecological, and technological systems in the Anthropocene. *Npj Urban Sustainability, 3*(1), Article 1. https://doi.org/10.1038/s42949-023-00120-1

Chu, E., Anguelovski, I., & Carmin, J. (2015). Inclusive approaches to urban climate adaptation planning and implementation in the Global South. *Climate Policy, 16*(3), 372–392. https://doi.org/10.1080/14693062.2015.1019822

Collier, M. J., Frantzeskaki, N., Connop, S., Dick, G., Dumitru, A., Dziubała, A., Fletcher, I., Georgiou, P., Hölscher, K., Kooijman, E., Lodder, M., Madajczyk, N., McQuaid, S., Nash, C., Osipiuk, A., Quartier, M., Reil, A., Rhodes, M.-L., Rizzi, D., & Xidous, D. (2023). An integrated process for planning, delivery, and stewardship of urban nature-based solutions: The Connecting Nature Framework. *Nature-Based Solutions, 3*, Article 100060. https://doi.org/10.1016/j.nbsj.2023.100060

Crowley, K., Jackson, R., O'Connell, S., Karunarthna, D., Anantasari, E., Retnowati, A., & Niemand, D. (2022). Cultural heritage and risk assessments: Gaps, challenges, and future research directions for the inclusion of heritage within climate change adaptation and disaster management. *Climate Resilience and Sustainability, 1*, Article e45. https://doi.org/10.1002/cli2.45

Cumpston, Z. (2020). Cities are country: Illuminating aboriginal perspectives of biodiversity in urban environments. Research synthesis. *Clean Air and Urban Landscapes Hub*. https://doi.org/10.5281/zenodo.4782538

Cumpston, Z., Fletcher, M.-S., & Head, L. (2022). *Plants: Past, present and future*. Thames and Hudson.

Elmqvist, T., Fragkias, M. (Ed.), Goodness, J. (Ed.), Güneralp, B., Marcotullio, P. J. (Ed.), McDonald, R. I. (Ed.), Parnell, S. (Ed.), Schewenius, M., Sendstad, M., Seto, K. C., & Wilkinson, C. (2013). *Urbanization, biodiversity and ecosystem services: Challenges and opportunities*. Springer Nature. https://doi.org/10.1007/978-94-007-7088-1

English, P. B., Richardson, M. J., & Garzón-Galvis, C. (2018). From crowdsourcing to extreme citizen science: Participatory research for environmental health. *Annual Review of Public Health, 39*(1), 335–350.

Fanning, A. L., & Raworth, K. (2025). Doughnut of social and planetary boundaries monitors a world out of balance. *Nature, 646*(8083), 47–56. https://doi.org/10.1038/s41586-025-09385-1

Frantzeskaki, N., & McPhearson, T. (2022). Mainstream nature-based solutions for urban climate resilience. *BioScience, 72*(2), 113–115. https://doi.org/10.1093/biosci/biab105

Frantzeskaki, N., Lambertucci, S., Carr, E., Dempsey, J., Karamidehkordi, E., Sengupta, A., Rojas Marchini, F., Vogel, C., Andriamahefazafy, M., Boonstra, W., Espinoza-Cisneros, E., Garcia, K., Morita, K., Nelson, V., Ojeda, D.,

Plieninger, T., Stirling, A., Tokunaga, K., Chen, R., ... & Guibal, C. (2024). Overcoming the challenges of achieving transformative change towards a sustainable world. In K. O'Brien, L. Garibaldi, & A. Agrawal (Eds.), *Thematic assessment report on the underlying causes of biodiversity loss and the determinants of transformative change and options for achieving the 2050 Vision for Biodiversity of the Intergovernmental Science-Policy Platform on Biodiversity and Ecosystem Services*. IPBES secretariat. https://doi.org/10.5281/zenodo.11382246

Frantzeskaki, N., McPhearson, T., & Kabisch, N. (2021). Urban sustainability science: Prospects for innovations through a system's perspective, relational and transformations' approaches. *Ambio, 1–9*. https://doi.org/10.1007/s13280-021-01521-1

Gazing Wolf, J., Ignace, D. D., David-Chavez, D. M., Jennings, L. L., Smiles, D., Blanchard, P., Simmons, E., Doan-Crider, D., Kills, R. P. S.-S., Montgomery, M., Nelson, M. K., Black Elk, L., Black Elk, L., Bridge, G., Chischilly, A. M., Deer, K., DeerinWater, K., Ecoffey, T., Vergun, J.,..., & Rattling Leaf, J. (2024). Centering indigenous knowledges in ecology and beyond. *Frontiers in Ecology and the Environment*, *22*(7), e2776. https://doi.org/10.1002/fee.2776

Gulsrud, N. M., Hertzog, K., & Shears, I. (2018). Innovative urban forestry governance in Melbourne? Investigating "green placemaking" as a nature-based solution. *Environmental Research, 161*, 158–167. https://doi.org/10.1016/j.envres.2017.11.005

Hurlimann, A., Moosavi, S., Warren-Myers, G., March, A., Bush, J., & Cutroni, L. (2025). Facilitating climate change action across built environment life stages—Perspectives from built environment professionals. *Cities, 165*, Article 106113. https://doi.org/10.1016/j.cities.2025.106113

IPBES. (2019). *Global assessment report on biodiversity and ecosystem services of the intergovernmental science-policy platform on biodiversity and ecosystem services*. IPBES secretariat. https://doi.org/10.5281/zenodo.4782538

Irvin, A., Oke, C., Khan, S., Jance IV, B., Ranalder, L., Barth, B., Deacon, A., Abusamra, R., Addison, A., Adebisi, J., Adsett, C., Ailio, E., Al Jadaa, D., Aslan, S., Assalini, S., Assarkhaniki, Z., Astolfo, J. P., Badino, M., Bai, X., ..., & Williams, A. (2024). Next-generation city climate action: Findings from the 2024 Innovate4Cities conference and update to the global research and action agenda on cities and climate change science (GRAA). Global Covenant of Mayors for Climate and Energy, UN-Habitat & University of Melbourne. https://doi.org/10.26188/27147612

Kendal, D., Egerer, M., Byrne, J. A., Jones, P. J., Marsh, P., Threlfall, C. G., Allegretto, G., Kaplan, H., Nguyen, H. K. D., Pearson, S., Wright, A., & Flies, E. J. (2020). City-size bias in knowledge on the effects of urban nature on people and biodiversity. *Environmental Research Letters, 15*(12), Article 124035. https://doi.org/10.1088/1748-9326/abc5e4

Mansur, A. V., McDonald, R. I., Güneralp, B., et al. (2022). Nature futures for the urban century: Integrating multiple values into urban management. *Environmental Science & Policy, 131*, 46–56. https://doi.org/10.1016/j.envsci.2022.01.013

McPhearson, T., Andersson, E., Grilo, F., Lopez, B., Zein, N., Grimm, N., Wellmann, T., Kendal, D., Haase, D., & Lausch, A. (2023). Urban ecological resilience: Ensuring urban ecosystems can provide nature-based solutions. In T. McPhearson, N. Kabisch, & Frantzeskaki, N. (Eds.), *Nature-based solutions for cities* (pp. 49–81), Edward Elgar Publishing. https://doi.org/10.5281/zenodo.4782538

McPhearson, T., Cook, E. M., Berbés-Blázquez, M., Cheng, C., Grimm, N. B., Andersson, E., Barbosa, O., Chandler, D. G., Chang, H., Chester, M. V., Childers, D. L., Elser, S. R., Frantzeskaki, N., Grabowski, Z., Groffman, P., Hale, R. L., Iwaniec, D. M., Kabisch, N., Kennedy, C., & Troxler, T. G. (2022). A social-ecological-technological systems framework for urban ecosystem services. *One Earth, 5*(5), 505–518. https://doi.org/10.1016/j.oneear.2022.04.007

McPhearson, T., Frantzeskaki, N., Ossola, A., Diep, L., Anderson, P. M. L., Blatch, T., Collier, M. J., Cook, E. M., Culwick Fatti, C., Grabowski, Z. J., Grimm, N. B., Haase, D., Herreros-Cantis, P., Kavonic, J., Lin, B. B., Lopez Meneses, D. H., Matsler, A. M., Moglia, M., Morató, J.,..., & Zhou, W. (2025). Global synthesis and regional insights for mainstreaming urban nature-based solutions. *Proceedings of the National Academy of Sciences, 122*(29), e2315910121. https://doi.org/10.1073/pnas.2315910121

McPhearson, T., Kabisch, N., & Frantzeskaki, N. (2023). Towards mainstreaming nature-based solutions for achieving biodiverse, resilient, and inclusive cities. In *Nature-based solutions for Cities* (pp. 363–375). Edward Elgar Publishing.

McPhearson, T., Pickett, S. T. A., Grimm, N. B., Niemelä, J., Alberti, M., Elmqvist, T., Weber, C., Haase, D., Breuste, J., & Qureshi, S. (2016). Advancing urban ecology toward a science of cities. *BioScience, 66*(3), 198–212. https://doi.org/10.1093/biosci/biw002

McPhearson, T. E. et al. (Eds.) (2021). In *Resilient urban futures* (pp. 173–186). Springer International Publishing. https://doi.org/10.1007/978-3-030-63131-4_12

McPhearson, T. E. et al. (2022). A social-ecological-technological systems approach to urban ecosystem services. *One Earth, 5*(5), 505–518. https://doi.org/10.1016/j.oneear.2022.04,.007

Mikaelsson, M. A., & Lager, F. (2024), *Justice and equity in climate change adaptation: Overview of an emerging agenda*. Stockholm Environment Institute. https://doi.org/10.5281/zenodo.4782538

Nelson, M. K., & Shilling, D. (Eds.). (2018). *Traditional ecological knowledge: Learning from Indigenous practices for environmental sustainability*. Cambridge University Press.

O'Donnell, M., Collier, M., Pineda-Pinto, M., Cooper, C., Nulty, F., & Castañeda, N. R. (2025). Redefining co-design for social-ecological research and practice: A systematic literature review. *Environmental Science & Policy, 164*, Article 103998. https://doi.org/10.1016/j.envsci.2025.103998

O'Brien, K., Garibaldi, L. A., Agrawal, A., Bennett, E., Biggs, R., Calderón Contreras, R., Carr, E. R., Frantzeskaki, N., Gosnell, H., Gurung, J., Lambertucci, S. A., Leventon, J., Chuan, L., Reyes García, V., Shannon, L., Villasante, S., Wickson, F., Zinngrebe, Y., & Périanin, L. (2025). IPBES transformative change assessment: Summary for policymakers (10.2.0).

Orlove, B., Sherpa, P., Dawson, N., Adelekan, I., Alangui, W., Carmona, R., Coen, D., Nelson, M. K., Reyes-García, V., Rubis, J., Sanago, G., & Wilson, A. (2023). Placing diverse knowledge systems at the core of transformative climate research. *Ambio, 52*(9), 1431–1447. https://doi.org/10.1007/s13280-023-01857-w

Parmesan, C., Anshari, G., Buotte, P., Campbell, D., Castellanos, E., Cowie, A. et al. (2022). Nature-based solutions for climate change: Mitigation and adaptation. In H.-O. Pörtner, D. C. Roberts, M. Tignor, E. S. Poloczanska, K. Mintenbeck, A. Alegría et al. (Eds.), *Climate change 2022: Impacts, adaptation, and vulnerability. Contribution of working group II to the sixth assessment report of the Intergovernmental Panel on Climate Change* (pp. 303–309). Cambridge University Press.

Pereira, L. M., Davies, K. K., den Belder, E., Ferrier, S., Karlsson-Vinkhuyzen, S., Kim, H., Kuiper, J. J., Okayasu, S., Palomo, M. G., Pereira, H. M., Peterson, G., Sathyapalan, J., Schoolenberg, M., Alkemade, R., Carvalho Ribeiro, S., Greenaway, A., Hauck, J., King, N., Lazarova, T., & Lundquist, C. J. (2020). Developing multiscale and integrative nature–people scenarios using the nature futures framework. *People and Nature, 2*(4), 1172–1195. https://doi.org/10.1002/pan3.10146

Pörtner, H. O., Scholes, R. J., Agard, J., Archer, E., Arneth, A., Bai, X., Barnes, D., Burrows, M., Chan, L., Cheung, W. L., Diamond, S., Donatti, C., Duarte, C., Eisenhauer, N., Foden, W., Gasalla, M. A., Handa, C., Hickler, T., Hoegh-Guldberg, O.,..., & Ngo, H. T. (2021). *IPBES-IPCC co-sponsored workshop report on biodiversity and climate change*. IPCC, I. a. https://doi.org/10.5281/zenodo.4782538

Tang, C., & Bush, J. (2025). Cooler streets for a cycleable city: Assessing policy alignment. *Buildings and Cities, 6*(1), 1–13. https://doi.org/10.5334/bc.508

Tengö, M., Austin, B. J., Danielsen, F., & Fernández-Llamazares, A. (2021). Creating synergies between citizen science and Indigenous and local knowledge. *BioScience, 71*(5), 503–518. https://doi.org/10.1093/biosci/biab023

Tzoulas, K., Korpela, K., Venn, S., Yli-Pelkonen, V., Kaźmierczak, A., Niemela, J., & James, P. (2007). Promoting ecosystem and human health in urban areas using green infrastructure: A literature review. *Landscape and Urban Planning, 81*(3), 167–178. https://doi.org/10.1016/j.landurbplan.2007.02.001

Waterton, EE. (2015). Heritage and community engagement. In T. Ireland & J. Schofield (Eds.), *The ethics of cultural heritage* (pp. 53–67). Springer. https://doi.org/10.1007/978-1-4939-1649-8_4

West, S., Haider, L. J., Hertz, T., Mancilla Garcia, M., & Moore, M. L. (2024). Relational approaches to sustainability transformations: Walking together in a world of many worlds. *Ecosystems and People, 20*(1), Article 2370539. https://doi.org/10.1080/26395916.2024.2370539

Zhou, W., Pickett, S. T. A., & McPhearson, T. (2021). Conceptual frameworks facilitate integration for transdisciplinary urban science. *NPJ Urban Sustainability,1*(1), Article 1. https://doi.org/10.1038/s42949-020-00011-9

CHAPTER 10

Systems Approaches to City Research and Innovation: Infrastructure, Resource Efficiency, Public Services, and Health

Gian Carlo Delgado-Ramos, Hilda Blanco, Carlos Moreno, Mattia Federico Leone, Carlos Alberto González-Guzmán, Ana Rosa Moreno Sánchez, and Nicola Tollin

10.1 Introduction

Urban transformation is a multidimensional and multiscale process that emerges from the recognition of cities as complex and dynamic systems. It aims to transcend sectoral or siloed planning, design, and action as a

G. C. Delgado-Ramos (✉)
Institute of Geography, National Autonomous University of Mexico, Mexico City, Mexico
e-mail: giandelgado@unam.mx

H. Blanco
University of Washington, Seattle, WA, USA
e-mail: hblanco@uw.edu

C. Oke et al. (eds.), *A Global Research and Action Agenda on Cities and Climate Change Science*,
https://doi.org/10.1007/978-981-95-6517-7_10

prerequisite for enabling a comprehensive approach toward more sustainable, resilient, inclusive, just, and healthy cities (Raven et al., 2025a, 2025b).

As noted in the introduction to Sect. 10.3 of this book, urban transformation can be structured around a set of interconnected systemic elements within four transformation realms: land and ecosystems (the biophysical context), urban infrastructure and public services, urban resource use efficiency, and socioeconomic and governance structures.

This chapter focuses on seven systemic elements which relate to the "systems approach levels" outlined in the Global Research Action Agenda on Cities and Climate Science (GRAA). The following section addresses land and ecosystems, with emphasis on land uses and urban planning, as other related aspects have already been discussed in Chapter 9. Section 10.3 explores the realms of urban infrastructure, public services, and urban resource flows, by assessing opportunities to advance climate solutions. It also addresses the role of urban social infrastructure in connecting governance innovation and socioeconomic transformation. Further discussion on governance, partnerships, and finance is provided in

---

C. Moreno
IAE Paris Sorbonne, Université Paris 1 Panthéon-Sorbonne, Paris, France
e-mail: Carlos.Moreno@iae.pantheonsorbonne.fr

M. F. Leone
Department of Architecture, University Naples Federico II, Naples, Italy
e-mail: mattia.leone@unina.it

C. A. González-Guzmán
MetroCali S.A., Cali, Colombia
e-mail: cargonza6@gmail.com

A. R. Moreno Sánchez
Faculty of Medicine, National Autonomous University of Mexico (Retired), Mexico City, Mexico
e-mail: ana_rosa_moreno@yahoo.com

N. Tollin
UNESCO Chair on Urban Resilience, University of Southern Denmark, Odense, Denmark
e-mail: nto@iti.sdu.dk

Chapter 11. Other issues related to health, justice, equity, diverse knowledge systems, and co-production practices are included as cross-cutting aspects where relevant.

10.2 Land Uses

Today, most of the world's population lives in cities. In 2018, 55.3% of people resided in urbanized areas, a proportion expected to increase to 68.4% by 2050 (UN, 2019). According to the Intergovernmental Panel on Climate Change (IPCC), without major changes, urban inhabitants will experience the most severe consequences of rising temperatures this century (Lwasa et al., 2022). Major sources of greenhouse gas emissions and air pollution include industrial and commercial activities, motor vehicles, and residential energy use. While the third of these sources depends mostly on building design, construction materials, and household consumption practices, the first two are largely influenced by land-use patterns, and thus urban planning, as discussed below.

10.2.1 Land-Use Planning and Zoning at the Core of Urban-Planning Practice

Land-use planning is the major distinctive practice in urban planning (Barnett, 2012; Elliot, 2008; Hirst, 2014). The separation of land uses through zoning has long been a central urban-planning practice and a key enabler of industrialization in cities. By separating industrial and commercial areas from residential areas, zoning laws reduced residents' direct exposure to pollution. However, this approach also allowed polluting industries and other noxious uses to operate largely unchallenged until the rise of the environmental-protection movement in the mid-twentieth century.

The separation of industrial, commercial, and residential areas was driven by rising automobile ownership and truck-based freight, both major sources of urban air pollution. With the decline of large-scale twentieth-century industrial models and the rise of service economies and light manufacturing, which increasingly locate value creation within or near urban centers, proximity-based urban models have gained prominence as frameworks for reshaping urban form and functions based on low-carbon accessibility, citizen well-being, and polycentric development, with the overall aim of integrating mixed land uses and sustainable urban densities with physical systems to foster urban resilience (UNEP & UN-Habitat, 2021).

Polycentric and neighborhood-oriented city models decentralize urban functions across multiple centers, each providing access to essential services, employment, and cultural amenities. Since 2020, research across Europe, North America, Latin America, and Asia has documented the rapid diffusion of proximity-oriented planning principles (Table 10.1 gives examples).

Meta-analyses and systematic reviews reveal substantial methodological heterogeneity: cities vary in how they define proximity (as 10, 15, or 20 minutes), the range of services included, and the spatial and social metrics used (Carvalho et al., 2025; Bruno et al., 2024; Jin et al., 2024; Allam et al., 2024; DUT, 2024). Despite these variations, the concept rests today on a common foundation encompassing spatial-proximity assessments, well-being indicators, and public policies geared toward urban transitions.

These proximity-based urban models also comprise compactness, accessibility strategies, and service integration, as discussed further in Sect. 10.3. Notably, compactness extends beyond density to include livability, as exemplified by Copenhagen's Finger Plan and Portland's human-scale planning overlays, which promote pedestrian-friendly environments (Van Nes, 2021).

Compact urban design is characterized by short blocks, sustainable, dense building patterns, and interconnected street grids, all of which facilitate walking and cycling, reducing the use of private cars and related greenhouse gas emissions. Since it presupposes limiting urban sprawl, compact cities tend to have a low demand for vacant land, which helps preserve surrounding environmentally protected areas and urban green spaces with ecological value. When these spaces are integrated into a dense and continuous network of urban green and blue infrastructure connecting core areas and "steppingstones" through ecological corridors (Firehock & Walker, 2019; Forman, 1995; Gregory et al., 2021), this approach supports biodiversity and ecosystem health while improving urban and peri-urban environmental conditions. Such improvements can range from enhanced thermal comfort and reduced air and noise pollution (via CO_2 capture and particulate retention) to greater soil stability and water infiltration (Frantzeskaki, 2019; Kumar et al., 2019; Van Renterghem, 2019).

Table 10.1 Examples of proximity-oriented planning

City (country)	Study focus	Key findings
Melbourne (Australia)	"20-minute neighborhoods" through integrated planning	Integration of housing and mobility. Success in high-density areas, but suburban spread remains (Li et al., 2024; Rao et al., 2024)
Paris (France)	A city of proximity, diversity, density and ubiquity	Urban proximity with low-carbon mobility, urban forests, child-friendly streets, and local commerce, all contributing to reductions in greenhouse gas and air pollutants while improving overall quality of life and health (Christoforaki, 2024; Paris Government, 2022)
Thessaloniki (Greece)	Feasibility of compact neighborhood design	Strong community cohesion and mixed uses, but poor cycling infrastructure and lack of green space hinder implementation (Shoina et al., 2024)
Portland (USA)	"20-minute neighborhoods" integration to city's Climate Action Plan	Broad neighborhood mapping, investments in walking/cycling networks, and affordable housing near amenities (Rao et al., 2024; Funk & Glickman, 2023; C40, 2022)
Cedar Rapids (USA)	Integration of proximity goals into city's Climate Action Plan	Uses 15-minute accessibility measures to guide zoning and tree-cover equity programs (Funk & Glickman, 2023)
Bogotá (Colombia)	Mobility and walkability	Introduction of "barrios vitales" policy; increased cycle participation and local amenities (C40, 2022, 2023)
Shanghai (China)	Post-COVID planning experiments	Pilot projects achieved higher accessibility indices but struggled with population–service balance (Shanghai Municipal Government, 2024). Envisions 15-minute "community-life circles" as cells of the city connected to services and public space (Rao et al., 2024)
Hong Kong (China)	15-minute city index	Measures a 15-minute city index emphasizing accessibility to five urban functions: living, health care, education, entertainment, and public transit (Liu et al., 2024)
Utrecht (Netherlands)	Integration of 15-minute principles into transport and housing policy	Emphasis on active mobility and micro-mobility corridors; strong citizen-participation model (Knap et al., 2023)

Source Authors

Compactness is also an opportunity to promote place-based planning that reduces spatial inequalities and strengthens urban citizenship, as intended by France's "*quartiers prioritaires*" (Paris Government, 2022). In this context, proximity and accessibility to essential services—schools, health care, public transport, markets, and recreation—become defining features of sustainable and resilient neighborhoods. Concentrating services within well-connected, walkable areas allows polycentric cities to reduce reliance on long-distance travel and improve community resilience to climatic, economic, and infrastructural shocks. While this spatial logic enables urban functioning at lower environmental costs, its effective implementation depends on fostering inclusiveness, social cohesion, and place-based economic systems.

10.2.2 Reducing Pollution and Greenhouse Gas Emissions

This subsection explores five land-use and planning-related opportunities to reduce future urban pollution and GHG emissions: teleworking, shopping trips, recreational spaces, urban agriculture, and land-use patterns in existing cities.

Teleworking Increase

As internet access and artificial intelligence (AI) transform work, the need to separate land uses may decline, as remote work reduces demand for industrial and commercial zones and lowers commuting-related pollution.

The effect of the COVID-19 pandemic provided recent evidence of the potential expansion of remote work. In the USA, only 6.5% of private-sector employees worked from home in 2019, but this share increased during the pandemic across major industries: management, finance, insurance, information, and professional, scientific, and technical services. By 2021–2022, between 39 and 46% of workers in these sectors reported working remotely (Pabilonia & Redmond, 2024). Part-time teleworking is likely to increase among many workers, reducing commuting; a transition that will nonetheless take time, particularly in contexts affected by the digital divide. While full-time teleworking is feasible for certain occupations, it has often been associated with digital nomadism and its related effects, including gentrification and demand for larger homes (Mahadieva et al., 2025). The transition to teleworking should thus be carefully planned, particularly considering that increased home-based work may increase residential energy use and related CO_2 emissions, especially during peak hours (Mahadevia et al., 2025).

The growing use of personal electronic devices and, most critically, conventional air conditioning in buildings risks further straining centralized power systems (Shi et al., 2023; Simon & O'Brien, 2023). Mitigating these impacts requires accelerating the transition toward distributed energy systems, positioning cities as key prosumers in both the production and consumption of energy. Locally diversified energy communities and district-level energy hubs, which are designed to harness renewable resources alongside on-site storage solutions, can help balance demand and supply, reduce peak loads, and strengthen energy resilience amid growing risks of blackouts and network disruptions.

Shopping-Trip Decreases

Increasing internet shopping is likely to reduce vehicle trips, as it relies on centralized delivery systems. Estimates indicate that a single delivery truck can save hundreds of consumer trips; for example, in the USA one last-mile delivery truck can serve 200–300 packages in a single route, replacing around 180–200 in-person individual shopping trips (Steer & Fourth Economy, 2024).

However, the environmental and spatial implications of this shift are complex and may undermine goals of sustainable and resilient urban districts, because e-commerce demand intensifies last-mile deliveries, congestion, noise, and emissions, particularly in dense residential areas. In parallel, the expansion of logistics hubs, fulfillment centers, and data centers places additional pressure on peri-urban and natural land (Monstadt & Saltzman, 2025). These facilities typically require large, single-use parcels with limited integration into the urban fabric, threatening efforts to preserve or restore (peri)urban ecosystems. Moreover, the dominance of online retail can weaken local economies essential to the liveliness, cohesion, and walkability of compact and mixed-use neighborhoods (Viu-Roig, 2020; Kerr et al., 2024; Andruetto et al., 2024). To address these issues, a new supply model must be conceived in which logistics and fulfillment infrastructures evolve from isolated, land-intensive back-end facilities into a metropolitan-to-neighborhood network of multifunctional hubs. These hubs, which would be spatially and functionally embedded within the urban fabric, could be connected to smart waste-management systems to foster circularity and expand urban economic opportunities by hosting recycled material design, digital and distributed manufacturing, start-up incubation, and vocational education, thus reconnecting logistics into physical, economic, and social realms of cities in line with polycentric- and sustainable-development principles (Elwakil et al., 2023; ITF-OECD, 2024).

Meeting and Recreational Places

As remote work increases, meeting and recreational spaces will gain importance, and thus should be equitably integrated across residential areas. These spaces can foster sustainable behaviors by facilitating everyday interactions, strengthening proximity, reducing motorized travel and emissions, and enhancing environmental awareness. Their effectiveness relies on intentional design, diversity of uses, and integration within the surrounding community context to reinforce sustainability objectives, social organization, and solidarity (European Commission, 2024).

Renewed attention to public spaces and multifunctional community hubs is likewise critical to anchoring climate-adaptation strategies in local priorities, ensuring that transformative change remains meaningful, equitable, and rooted in everyday urban life. In this context, neighborhood-scale interventions provide key opportunities to test innovative concepts and approaches that bridge diverse knowledge systems and community aspirations (Raven et al., 2025a, 2025b).

Sustainable Urban Agriculture

Sustainable urban agriculture should be actively promoted, as it can enhance community health and well-being while supporting greener, more livable urban spaces and more resilient local economies. Sustainable and healthy food can be produced not only in peri-urban areas but also where industrial and vehicle pollution has declined. Sustainable urban agriculture should be considered as an integral component of broader green–blue infrastructure networks. When designed in synergy with ecological connectivity strategies, community gardens, rooftop farms, and small-scale cultivated spaces can function as ecological "stepping-stones" or corridors within polycentric urban and metropolitan systems, contributing to biodiversity conservation, microclimatic regulation, and the adaptive capacity of cities (De Zeeuw & Drechsel, 2015).

From this perspective, opportunities for scaling up include off-soil and vertical farming, floating agriculture, and agrivoltaic systems. These approaches can be embedded into underused or hybrid urban spaces to promote local circular supply chains and foster regenerative urban metabolisms, where both food and non-food species can be transformed into by-products such as insulation materials, bio-based composites, and other sustainable goods (Ackerman et al., 2012).

Changing Land Use Patterns in Existing Cities

Existing urban land-use patterns must be restructured if greenhouse gas and air-pollutant emissions from major sources—particularly industry and vehicle fleets—are to be significantly reduced. The separation of land use increases travel distances and the number of trips required for work, education, shopping, or recreation; each with energy, material, and pollution costs that depend on the type of transport mode (with motorized private vehicles the least efficient). Mixed uses, by contrast, facilitate walking and cycling, improving the health of individuals, cities, and the planet while fostering continuous public awareness of local air quality.

As economies and built environments transition toward non-polluting energy sources, a greater mix of land uses could foster healthier and more vibrant urban environments by promoting walking, cycling, and other forms of active mobility between diverse destinations, thereby reducing reliance on motorized travel. Proximity—a central principle of this transition—aims to reduce energy and material use and related emissions while enhancing well-being by aiming for cities where most daily needs can be met within a short walk or bike ride from home (Moreno et al., 2021). Reaching these goals requires more than spatial reconfiguration; it demands a transformative shift in how urban development is conceived and governed. Land-use planning must evolve from a static, regulatory function into a dynamic, adaptive practice that recognizes cities as complex systems shaped by feedback loops, interdependencies, and systemic risks. Urban form, mobility, energy, housing, and consumption patterns are deeply interconnected drivers of carbon emissions. Transforming them demands integrated strategies that reflect city-specific conditions, including demographic dynamics, economic structures, and ecosystem features (Raven et al., 2025a, 2025b). Aligned with systems thinking and long-term resilience principles, this planning perspective promotes decision-making processes that navigate uncertainty, learn from disruption, and steer structural transformation by fostering social innovation and realigning urban economies to operate within planetary boundaries (Folke et al., 2021; Frantzeskaki et al., 2018).

10.3 Urban Infrastructure, Circularity, and Public Services.

10.3.1 Sustainable Buildings

Buildings account for about 40% of global primary energy consumption and 24% of global greenhouse gas emissions (IRP-UNEP, 2018), making them central to sustainability and climate dynamics across multiple scales. At the macro scale, buildings contribute to soil sealing and both influence and are influenced by the albedo effect (Ouyang et al., 2022). Their distribution, size, and construction materials affect urban ventilation and thermal conditions (Nugroho et al., 2022), which in turn influence public health, as demonstrated during the COVID-19 pandemic (Mahadevia et al., 2025).

Commercial buildings play an important role in building-related carbon emissions, and thus in the development of mitigation strategies (Ghorbany et al., 2025). However, residential buildings are particularly important due to their dominant share of the built environment, resource demands, and associated greenhouse gas emissions, and strong influence on urban form (PNUMA, 2021).

At the micro scale, building design and construction not only influence residents' quality of life but also determine resource demand throughout buildings' entire lifespan, from construction to operation and maintenance (IRP-UNEP, 2018; PNUMA, 2021; Attia et al., 2024). This, in turn, creates path dependencies that affect the intensity of both embodied and operational carbon (Chen et al., 2023; Ekonomou & Menegaki, 2023; Ghorbany et al., 2025; Liyanage et al., 2024).

Solutions to improve building resource-use efficiency include bioclimatic design; the use of sustainable and locally sourced materials; changes in construction practices and the extension of building lifespans; the substitution of energy-efficient technologies for lighting, refrigeration, and climate control for conventional systems; the adoption of alternative ecotechnologies, including renewable energy and rainwater-capture systems; retrofitting to reduce thermal losses; and the reuse and recycling of construction materials and building components (IRP & UNEP, 2018; PNUMA, 2021; UNEP & UN-Habitat, 2021). Further positive impacts on urban resource efficiency can be achieved when buildings are paired with compact urban layouts and mixed land uses (IRP & UNEP, 2018; UNEP & UN-Habitat, 2021).

Building design and construction materials also influence—although not exclusively—indoor environmental health, including air quality, which became particularly relevant during the COVID-19 lockdowns (Jarrahi et al., 2024; Mahadevia et al., 2025; Megahed & Ghoneim, 2021). Indoor temperatures and air quality can be improved not only through ventilation and shading, but also by incorporating vegetation both indoors and in surrounding areas (Lin et al., 2023; Mangone et al., 2014). Similarly, the relationship between street surfaces and ambient temperatures—increasingly affected by extreme heat and the urban heat-island effect in many regions—warrants attention, as it influences outdoor and indoor environmental conditions (Amani et al., 2025; Wong et al., 2024). In both cases, further research is needed to deepen understanding of these effects across diverse urban contexts. The latter is of major relevance as extreme heat and wildfire smoke days increase, as ventilation and shading in many cities might be inadequate to protect people from both heat and poor air quality.

The way buildings are constructed also shapes their vulnerability and that of their occupants. Building codes can enhance resilience if they incorporate the growing impacts of climate change. In this context, it is important to note that housing is not merely a physical structure but a dynamic social infrastructure where interactions, proximity, services, and shared experiences converge to foster belonging, stability, and inclusion. Recognizing systemic impacts is crucial to address how housing and social infrastructure, including facilities for nonprofit social services and community activities, are increasingly displaced by real-estate speculation and development projects in large and growing cities; this displacement calls for equity-focused, neighborhood-scale strategies (UN-Habitat, 2020).

Cities can therefore address multiple objectives simultaneously through a comprehensive policy aimed at providing affordable, sustainable, and climate-ready housing, as well as public services and spaces that nurture learning, dialogue, and solidarity, such as libraries, parks, and cultural centers. This type of strategy can align local priorities, funding opportunities, and climate action.

10.3.2 Sustainable Mobility and Transport

Efforts to advance sustainable mobility should focus on modal share; that is, the percentage of trips made by each mode of transport, including walking and cycling, public transport, and private vehicles. The primary objective is to substantially increase the share of active and public transport relative to private motorized travel.

The desirable modal shift in mobility can be pursued through two main approaches: first, by optimizing the mixture and proximity of the strategic location of land uses across the territory as both origins and destinations; and second, by inducting mobility scenarios that promote the use of sustainable transport modes.

Sustainable accessibility supports this objective by prioritizing active and public transportation through investments in multimodal networks that efficiently connect neighborhoods while reducing reliance on private vehicles.

Curitiba and Vienna exemplify cities where high-quality bus and tram systems reduce mobility-based exclusion and contribute to air quality and carbon-neutrality goals (OECD, 2024). A similar example applies to cable-car systems aimed at enhancing territorial accessibility for low-income neighborhoods located on the urban periphery. Notable cases in Latin America include Metrocable in Medellin, TransMiCable in Bogotá, and MIO Cable in Cali, all in Colombia (Bocarejo et al., 2014; Cardona-Urrea et al., 2023); Mi Teleférico in La Paz, Bolivia (Martínez et al., 2024); and Cablebús in Mexico City, Mexico (Prieto-Rodriguez et al., 2025).

Despite these examples, it is important to recognize that modal choice, or the decision of which transport mode to use for a given trip, is strongly influenced by the spatial distance between origins and destinations. For example, trips of up to 2 km are more likely to be made on foot, those up to 5 km by bicycle, and those exceeding 5 km typically require motorized modes, whether public or private, fuel-based or electric.

Similarly, trip length is significantly influenced by two factors: at a macro scale; the overall size and structure of the city, and at a zonal scale; the diversity of land uses and activities that act as trip generators (origins, such as residential areas) and attractors (destinations such as those related to commerce, services, industry, institutions, and other locations of activity). Larger cities tend to generate longer trips in a greater proportion of cases, while lower land-use diversity at the zonal level increases the need for interzonal travel.

Although significant efforts have been made in urban planning to promote greater diversity of uses and activities at the zonal level and to contain urban sprawl, results have often been insufficient. It is unlikely that this trend will be significantly reversed in the short term—particularly in large cities of less developed countries—without deep, systemic, and transformative interventions.

Objectives that involve changing the spatial distribution of urban land uses or activities—both trip generators and trip attractors—face complex processes, influenced by land tenure, housing affordability, stakeholders' investment-location preferences, and other factors. These conditions ultimately constrain the potential for a short-term transformation, often extending the required timeframes over decades due to infrastructure lock-ins and entrenched vested interests.

Actions that align urban land use with transport infrastructure—such as transit-oriented density or transport-oriented development models—are crucial, particularly alongside efforts to promote active mobility and prioritize public transport as the main mode of motorized travel, ideally through electrification where feasible. This set of actions involves enhancing accessibility at the zonal level and improving public transport connectivity at the citywide or metropolitan scale.

Public transport, when aligned with land-use planning, should thus be understood as the backbone of more sustainable mobility and a key determinant for addressing climate change. Unfortunately, public transport systems in many cities continue to face financial challenges in sustaining their operations, partly due to a decline in ridership since the COVID-19 pandemic and the increased availability of cheaper private cars (Karunakaran et al., 2023; Lee et al., 2024; Mohammadi-Mavi et al., 2025). Therefore, decisive action is required to ensure the implementation and economic viability of public transport operations through service improvements, funding schemes, and governance mechanisms. At the same time, strong measures must be implemented to discourage the use of private motorized transport while significantly improving infrastructure for active mobility, including complete streets (Messier, 2025; Nelson & Hibberd, 2023) and bike lanes (Wang et al., 2025).

10.3.3 Green Infrastructure

Nature-based solutions, ecosystem-based adaptation, and green infrastructure play a critical role in enhancing urban sustainability and resilience. They contribute to air-pollution mitigation, flood-risk reduction, thermal comfort, and the conservation of urban biodiversity, while also providing a range of other ecological services (Hautamäki et al.,

2025; Langemeyer & Baró, 2021; Zannat et al., 2024; Zarei & Shahab, 2024). Chapter 9 discusses this further.

Moreover, green urban spaces promote physical activity, which is key to addressing obesity and improving overall health (WHO, 2025), and offer exposure to nature, which helps reduce stress, improve cognitive function, and foster a sense of connection to the environment (Houlden et al., 2018; Keniger et al., 2013). These spaces are also essential for reducing health inequalities and strengthening community ties by restoring city-scale socio-ecological functions that have been eroded by urban sprawl, fragmentation, and the dominance of gray infrastructure (Kabisch et al., 2023; Vranayova et al., 2024).

Given the complex interactions between urbanization and public health, strengthening research in urban socio-ecological-technological systems (discussed in Chapter 9) is essential. Such research can inform health policies, infrastructure planning, and the provision of urban services, helping to bridge the gap between population needs and the health sector's capacity to implement preventive measures and respond effectively.

A promising approach lies in integrating green–blue infrastructure with investments in active mobility and public safety to deliver multipurpose, collaborative outcomes. Some cities are already moving in this direction. For instance, Mexico City's green infrastructure strategy outlines a deployment roadmap through 2050, aligned with active-mobility initiatives and public space revitalization (EVALUA, 2024). Similarly, Stockholm's green–blue infrastructure strategy emphasizes the integration of functioning ecosystems, green spaces, and nature-based solutions and ecosystem-based adaptation into long-term sustainable and resilient urban planning (City of Stockholm, 2024).

Further research is still needed to deepen understanding of the health and well-being benefits associated with green and blue urban spaces; research on the causal relationship between environmental "greenness" and "blueness" and adult mental health outcomes is particularly vital (Gascon et al., 2015). There is also a pressing need to assess how incorporating these benefits into urban planning could inform more equitable and evidence-based decision-making, especially in cities where such considerations remain underdeveloped.

Importantly, the advancement of nature-based solutions should prioritize traditionally marginalized and vulnerable communities, who often experience both distinct benefits and disproportionate disadvantages from such interventions; the latter include the risk of displacement through green gentrification (Bressane et al., 2024; Cucca et al., 2023). Finally, greater attention must be given to how climate change may affect the long-term viability and performance of urban green–blue infrastructure (Reynolds et al., 2020), and therefore to the adjustments needed in its planning and management.

10.3.4 Resource-Use Efficiency and Circular Economies

Cities account for over 75% of global material-resource consumption and, consequently, a substantial share of embodied emissions. Under a business-as-usual scenario, global urban resource consumption is projected to reach approximately 90 billion tons—a 50–60% increase compared to the combined rural and urban material and energy consumption in the year 2000 (IRP-UNEP, 2018).

While some urban material flows are short-lived, others, particularly those embedded in buildings and infrastructure, can persist for decades, constituting most of the urban material stock (UNEP, 2021). Despite the longer lifespan of materials used in buildings and infrastructure, the construction sector remains one of the most resource-intensive and metabolically dominant components of urban systems (Li et al., 2019).

To transition toward more efficient and sustainable urban systems, it is essential to shift from a linear to a circular approach to resource use. This means closing material loops, including those for water and biomass (like food), through strategies such as reducing, reusing, repairing, refurbishing, remanufacturing, repurposing, recycling, and recovering resources, or by harnessing energy from urban outflows like waste and wastewater. These efforts must be accompanied by substantial reductions in overall resource consumption driven by significant behavioral change and grounded in principles of social justice.

Some cities have begun implementing circular-economy policies, yet most current efforts address only a small fraction of urban material outflows, and are largely driven by private-sector initiatives, including those on industrial symbiosis (Neves et al., 2019; Ruini et al., 2025).

For cities to successfully advance circularity—which must be understood as extending far beyond waste management and recycling—the development of specific, coherent, and multi-level regulatory frameworks is essential. These frameworks require alignment with updated urban-planning instruments, building codes, and fiscal tools such as subsidies (Williams, 2022), along with other normative mechanisms for key processes that shape urban resource use, such as end-of-life regulations for consumer and industrial products (UNEP & UN-Habitat, 2021). An example of this type of alignment is Amsterdam's policy on circular tendering and land allocation, which applies to urban transformation and infrastructure renovation and demolition (Williams, 2022).

Fully harnessing the potential of urban circularity requires fundamental changes in business, industry, and labor markets. To be socially meaningful, these changes must be carefully guided by principles of human prosperity, equity, and justice (UNEP & UN-Habitat, 2021).

Urban circularity also involves the development of urban metabolic models and systematic data collection: drawing on material and energy flow analyses and life-cycle assessments to inform decision-making processes and enable accountability for actions and outcomes (UNEP & UN-Habitat, 2021).

10.3.5 Public Services

Inclusive and resilient urban infrastructures are crucial for ensuring the sustainable provision of essential services and for supporting social and ecological transformation (OECD, 2021). Basic urban infrastructure, such as transportation, energy, water, sanitation, and waste management, remains the foundational layer for public health and equity. Its planning, therefore, must integrate principles of climate resilience, affordability, and environmental justice. These services, which are supported by urban infrastructure, can be improved through technological innovation and the advancement of integrated citizen-centered service models.

Spatially defined zones for service provision or urban service areas can improve service quality, efficiency, and inclusivity. Beyond functional provision, the spatial configuration of public services is critical in fostering equitable and sustainable urban living. When embedded within compact, well-connected, mixed-use districts, public services become more accessible and responsive to local needs, reducing time and energy costs for users while increasing community resilience. This spatial logic supports the development of distributed networks of multifunctional "community hubs"—facilities that translate into physical spaces the needed transformative system change—calling planners and designers to reinvent urban services linked to mobility, energy, food distribution, health care, circular waste management, manufacturing, and education as novel hybrid spaces catalyzing urban community development (Raven et al., 2025b).

In the conventional literature, "community hubs" or "resilience hubs" are typically understood as neighborhood facilities that co-locate social, health, educational, and cultural services and act as centers of community interaction (McShane & Coffey, 2022). The integration of additional urban services, tailored to neighborhood-scale demand, can enhance local autonomy, reduce reliance on centralized infrastructure, and serve as anchors of polycentric development. In this emerging model, the proximity of diverse public services is not only a matter of convenience but a structural condition for inclusive decarbonization.

For example, the USDN "Resilience Hub" framework extends these facilities by embedding hybrid resilient energy systems (solar, storage, islanding) to ensure continuity during outages (USDN, 2019). This model can be further expanded, moving from passive service anchors to urban-services hubs that integrate intelligent waste collection and sorting, localized manufacturing or repair labs, logistics consolidation and distribution, and vocational education.

Integrated and Citizen-Focused Service Models

A sustainable urban system depends not only on the design and delivery of services but also on the shift from top-down to networked, citizen-focused models, which is essential for building systemic resilience.

The interoperability of services and systems is crucial to ensure seamless coordination across urban functions, enabling integrated responses to complex challenges and fostering more efficient, equitable, and resilient city operations.

Integrated services overcome institutional silos. Seoul's Community Service Centers or Medellín's Unidades de Vida Articulada are examples of interdependent facilities that offer education, culture, health, and recreation under one roof, enhancing administrative coherence and public trust, while reducing forced mobility, and thus CO_2-related emissions.

Aligning public-service delivery with the lived experiences of residents is also important and achievable in this context through citizen-centric services. "Care blocks" in Bogotá or Montreal's neighborhood "care houses" show how attention to proximity can enhance both dignity and equity in service access. Similarly, Mexico City's "Utopias" model has been developed to provide citizen-centric services under one roof in disadvantaged boroughs. Within the Utopias buildings—"Units of

Transformation and Organization for Inclusion and Social Harmony"—residents can access culture, health, sport, education, and training services (CDMX Government, 2024). These public spaces, expected to reach 100 by 2030, also serve as emergency shelters and form the core of a mid-term strategy to promote the development of urban subcenters and sub-subcenters connected through public transportation and active-mobility infrastructure.

Technology for Improved Climate Services and Urban Services

Effective climate action requires context-specific plans that respond to changing local climate conditions and evolving associated risks. AI, information technologies (IT) and big data can support and enhance such planning processes through climate services and complex simulations that account for land-use patterns, transport systems, resource use, economic profiles, and implementation capacities, while also considering how technologies reshape workplaces, consumption and travel behavior, and even planning process themselves.

Some cities are already developing climate-prediction strategies linked to rapid-implementation action plans that integrate IT, AI, and big data. For instance, Singapore (ETHETH zürich, no dated), Belo Horizonte (Government of Belo Horizonte, 2025), and 14 cities in the USA (Kempe, 2024) are exploring AI-assisted multi-scale microclimate models to identify drivers of urban climatic risks such as heat and flooding, simulate tendential scenarios, and generate design recommendations.

Models and simulations of local/regional climate change, its projected impacts, the potentiality of different actions, and alternative policy and governance scenarios (particularly those adopting a systemic approach to urban planning) can strengthen public understanding, build support for climate action, promote the interaction and integrated use of novel technologies, and foster collaboration across knowledge systems to co-generate more-effective local solutions.

Technological innovation can further enhance real-time responsiveness, inclusion, and sustainability in public-service provision; for example, through traffic regulation, energy grids, or waste management.

Participatory budgeting platforms, smart mobility dashboards, and AI-assisted utilities optimization illustrate how innovation can improve service quality. However, ensuring digital equity remains essential to prevent new forms of exclusion or technological dependency. The deployment of such technologies must be guided by ethical principles that address the digital divide, surveillance, consent, and data ownership to generate and sustain public trust, while also safeguarding against cybersecurity risks and the generation of misinformation intended to distort climate governance.

Moreover, the extent to which technological solutions may be adopted must respond to local priorities and urgent needs, placing human rights at the forefront of decision-making.

Also, in the Global South, such technology-based solutions should follow, or be implemented alongside without displacing, more accessible, affordable, and context-sensitive solutions grounded in Local and Indigenous Knowledges and practices.

10.3.6 Social Infrastructure for Connecting Governance Innovation and Urban Transformation

Social infrastructure encompasses the physical, institutional, and relational frameworks that support collective well-being and foster a sense of belonging (Hollis et al., 2023; Renner et al., 2024). It is also closely tied to the equitable distribution of climate investments and their expected outcomes (UNEP & UN-Habitat, 2021), because well-developed social infrastructure determines who can access and benefit from climate initiatives.

Communities with robust networks and social spaces tend to mobilize more resources and have greater influence over project outcomes, whereas less-connected or underserved areas often face barriers to engagement and equity. Strong social infrastructure builds trust and supports civic participation, both essential for scaling effective climate strategies and ensuring fair distribution of benefits. This interconnectedness extends further to the emergence of novel, agile, and inclusive forms of collaboration, including those involving city networks, enabling cities to collectively advance sustainability and resilience (Solecki et al., 2021).

The effectiveness of social infrastructure depends fundamentally on the socioeconomic and governance structures that frame it. Economic inequalities influence access to resources and shape 'individuals' and 'communities' capacity to engage in collective climate action. Governance systems—whether centralized or polycentric—define how responsibilities are distributed across public, private, and civic actors and how accountability is maintained (Bulkeley et al., 2017). Cities governed through transparent, participatory, and multi-level arrangements can align social infrastructure with inclusive and sustainable outcomes, while hierarchical or fragmented models have often perpetuated disparities, policy silos, and the transfer of risk (Dodman et al., 2022). In this sense, the governance of proximity—the coordination of mobility, housing, and services within urban territories—becomes an essential mechanism for linking social equity, spatial justice, and environmental performance.

Social infrastructure can thus serve as a connective tissue between governance innovation and socioeconomic transformation, translating climate and environmental agendas into tangible urban outcomes (see Chapter 4). Achieving this requires governance systems that integrate social and environmental policy domains, promote cross-sectoral learning, and empower citizens through participatory and deliberative processes (Schmitt & Wiechmann, 2018).

Public spaces and neighborhood hubs, but also educational and civic assets, function as local governance interfaces, given their role as platforms for citizen awareness, education, outreach, democratic expression, social solidarity, behavioral change, and neighborhood-level governance. Examples such as Amsterdam's Vrijplaats and Bologna's Civic Imagination Offices (Hollis et al., 2023) illustrate how such infrastructures can anchor democratic life while nurturing the social foundations for collective climate resilience (Colding et al., 2022).

10.4 Conclusion

Framed within a systemic approach, this chapter has examined key systemic elements shaping sustainable, resilient, and carbon–neutral urban transformation across four interdependent realms. Specific attention has been given to land use and urban planning, infrastructure and services, resource flows and efficiency, and the role of social infrastructure in connecting governance innovation and socioeconomic transformation. Within these transformation realms we demonstrate the value of mapping data and incorporating diverse knowledge sources across the analyzed interrelated systemic elements in the context of impacts on justice and equity.

The review has shown how land-use planning and zoning can steer cities toward adaptation and mitigation by integrating ecosystem services, conservation, and health co-benefits, linked across interrelated systemic elements (or systems approach levels of the GRAA). It highlights the transformative role of infrastructure and public services in delivering inclusive, low-carbon, and efficient urban functions. It also underscores the importance of circular resource flows and sustainable production and consumption through renewable energy, resource efficiency, and circular-economy practices. Finally, governance structures were identified as enablers of policy coherence, capacity-building, and knowledge coproduction, ensuring that climate action is guided by justice, equity, and responsiveness to local priorities.

Taken together, these systemic elements form an integrated framework for accelerating urban climate action and resilience, positioning cities as key actors in advancing regenerative, inclusive, and carbon–neutral futures.

Acknowledgements The authors acknowledge Chiara Morfeo, of C40 Cities, for thoughtfully reviewing an earlier draft version of this chapter.

Competing Interests

The authors have no conflicts of interest to declare.

References

Ackerman, K., Plunz, R., Conard, M., Katz, R., Dahlgren, E., & Culligan, P. (2012). *The potential for urban agriculture in New York City. Growing capacity, food security, and green infrastructure*. Columbia University. http://urbandesignlab.columbia.edu/files/2015/04/4_urban_agriculture_nyc.pdf

Allam, Z., Khavarian-Garmsir, A. R., Lassaube, U., Chabaud, D., & Moreno, C. (2024). Mapping the implementation practices of the 15-minute city. *Smart Cities*, *7*(4), 2094–2109. https://doi.org/10.3390/smartcities7040083

Amani, M. J., Tanzadeh, R., Nejad, F. M., Kabiri Nasrabad, M. M., Chalabii, J., & Rad, M. M. (2025). Urban sustainability through pavement technologies: Reducing urban heat islands with cool pavements. *Buildings*, *15*(3), 504. https://doi.org/10.3390/buildings15030504

Andruetto, C., Stenemo, E., & Pernestål, A. (2024). Towards sustainable urban logistics: Exploring the implementation of city hubs through system dynamics. *Transportation Research Interdisciplinary Perspectives, 27*, Article 101204. https://doi.org/10.1016/j.trip.2024.101204

Attia, M. M., Elboshy, B., Mohamed, A. S., Eita, M. A., & Shakra, K. E. (2024). Improving the energy efficiency performance of residential buildings in hot arid regions using lightweight concrete bricks. *Case Studies in Construction Materials, 21*, Article e03474. https://doi.org/10.1016/j.cscm.2024.e03474

Barnett, J. (2012). How codes shaped development in the United States, and why they should be changed. In S. Marshall (Ed.), *Urban coding and planning*. Taylor and Francis. https://doi.org/10.4324/9780203717561

Bocarejo, J. P., Portilla, I. J., Velasquez, J. M., Cruz, M. N., Peña, A., & Oviedo, D. R. (2014). An innovative transit system and its impact on low-income users: The case of the Metrocable in Medellín. *Journal of Transport Geography, 39*, 49–61. https://doi.org/10.1016/j.jtrangeo.2014.06.018

Bressane, A., Cunha Pinto, J. P., & Castro Medeiros, L. C. (2024). Countering the effects of urban green gentrification through nature-based solutions: A scoping review. *Natured-Based Solutions, 100131*. https://doi.org/10.1016/j.nbsj.2024.100131

Bruno, M., Monteiro Melo, H. P., Campanelli, B., & Loreto, V. (2024). A universal framework for inclusive 15-minute cities. *Nature Cities, 1*, 633–641. https://doi.org/10.1038/s44284-024.00119-4

Bulkeley, H., Coenen, L., Frantzeskaki, N., Hartmann, C., Kronsell, A., Mai, L., Marvin, S., McCormick, K., Van Steenbergen, F., & Voytenko Palgan, Y. (2017). Urban living labs: Governing urban sustainability transitions. *Current Opinion in Environmental Sustainability, 22*, 13–17. https://doi.org/10.1016/j.cosust.2017.02.003

C40. (2022). *Benchmark: 15-minute cities*. C40-Buenos Aires Ciudad. https://c40.my.salesforce.com/sfc/p/#36000001Enhz/a/1Q0000001rYk/VV8Uz59zv7yhjLcvL.sZ376T0fFvqac6zfu5o1fAizw

C40. (2023). *C40 cities annual report 2023*. https://www.c40.org/wp-content/uploads/2024/05/C40_annual_report_2023_V10.pdf

Cardona-Urrea, S., Soza-Parra, J., & Ettema, D. (2023). Aerial cable cars as a transit mode: A review of technological advances, service area characteristics, and societal impacts in Latin America and the Caribbean. *Transport Reviews, 44*(3), 684–708. https://doi.org/10.1080/01441647.2023.2294752

Carvalho, T., Farber, S., Manaugh, K., & El-Geneidy, A. (2025). Assessing the readiness for 15-minute cities: A literature review on performance metrics and implementation challenges worldwide. *Transport Reviews, 45*(5), 801–827. https://doi.org/10.1080/01441647.2025.2513530

Chen, Y., Ren, Z., Peng, Z., Yang, J., Chen, Z., & Deng, Z. (2023). Impacts of climate change and building energy efficiency improvement on city-scale building energy consumption. *Journal of Building Engineering, 78*, Article 107646. https://doi.org/10.1016/j.jobe.2023.107646

CDMX Government. (2024, December 31). Aviso por el que se dan a conocer las reglas de operación del Programa Social "Ciudad de Utopías, 2025". *Gaceta Oficial de la Ciudad de México*. https://www.transparencia.cdmx.gob.mx/storage/app/uploads/public/67a/3d2/c0e/67a3d2c0e6fa9894476571.pdf

Christoforaki, K. (2024). Towards a sustainable urban future: The 15-minute city approach in Paris, France. *Journal of Sustainable Development, Culture and Traditions, 24*(2024), 33–43. https://doi.org/10.26341/issn.2241-4002-2024-2a-3-T02045

City of Stockholm. (2024). *Environment programme 2030*. Stockholm, Sweeden. https://start.stockholm/globalassets/start/om-stockholms-stad/sa-arbetar-staden/miljo-och-klimat/environment-programme-2030.pdf

Colding, J., Bartherl, S., Ljung, R., Eriksson, F., & Sjöberg, S. (2022). Urban commons and collective action to address climate change. *Social Inclusion, 10*(1), 103–114. https://doi.org/10.17645/si.v10i1.4862

Cucca, R., Friesenecker, M., & Thaler, T. (2023). Green gentrification, social justice, and climate change in the literature: Conceptual origins and future directions. *Urban Planning, 18*(1), 283–295. https://doi.org/10.17645/up.v8i1.6129

De Zeeuw, H., & Drechsel, P. (2015). *Cities and agriculture*. Routledge.

DUT – Driving Urban Transitions. (2024). *Mapping of 15-minute city practices. Overview on strategies, policies and implementation in Europe and beyond*. https://dutpartnership.eu/sites/default/files/2025-07/DUT_15mC-Mapping_digital_final.pdf

Dodman, D., Hayward, B., Pelling, M., Castán Broto, V., Chow, W., Chu, E., Dawson, R., Khirfan, L., McPhearson, T., Prakash, A., Zheng, Y., Ziervogel, G., & Muñoz, T. A. (2022). Cities, settlements and key infrastructure. In *Climate change 2022: Impacts, adaptation and vulnerability. Contribution of working group II to the sixth assessment report of the IPCC*. Cambridge University Press.

Ekonomou, G., & Menegaki, A. N. (2023). The role of the energy use in buildings in front of climate change: Reviewing a system's challenging future. *Energies, 16*(17), 6308. https://doi.org/10.3390/en16176308

Elliot, D. L. (2008). *A better way to zone. Ten principles to create more livable cities*. Island Press.

Elwakil, R., Schroder, I., & Steemers, K. (2023). Circular maker cities: Maker space typologies and circular urban design. *Buildings, 13*(11), 2894. https://doi.org/10.3390/buildings13112894

ETH zürich. (no dated). *Cooling Singapore*. https://sec.ethz.ch/research/cs.html

European Commission. (2024). *Scoping the socio-economic performance of the EU proximity economy—Final report*. Publications Office of the European Union. https://data.europa.eu/doi/10.2826/1010

EVALUA. (2024). *Evaluación de la política ambiental de la Ciudad de México, 2018–2022*. EVALUA. www.evalua.cdmx.gob.mx/storage/app/media/2024/se/Informe%20medio%20ambiente%20EVALUA_final%20final%20ok_2.pdf

Firehock, K., & Walker, A. (2019). *Green infrastructure. Map and plan the natural world with GIS*. Esri Press.

Folke, C., Polasky, S., Rockström, J., Galaz, V., Westley, F., Lamont, M., Scheffer, M., Österblom, H., Carpenter, S. R., Stuart Chapin III, F., Seto, K. C., Weber, E. U., Crona, B. I., Daily, G. C., Dasgupta, P., Gaffney, O., Gordon, L. J., Hoff, H., Levin, S. A.,... & Walker, B. H. (2021). Our future in the Anthropocene biosphere. *Ambio*, *50*, 834–869. https://doi.org/10.1007/s13280-021-01544-8

Forman, R. T. (1995). *Land mosaics: The ecology of landscapes and regions.* Cambridge University Press.

Frantzeskaki, N. (2019). Seven lessons for planning nature-based solutions in cities. *Environmental Science & Policy, 93*, 101–111. https://doi.org/10.1016/j.envsci.2018.12.033

Frantzeskaki, N., Castán Broto, V., Coenen, L., & Loorbach, D. (2018). *Urban sustainability transitions*. Routledge.

Funk, K., & Glickman, J. (2023, June 13). Exploring the 15-minute city concept and its potential for communities of all sizes. *National League of Cities*. https://www.nlc.org/article/2023/06/13/exploring-the-15-minute-city-concept-and-its-potential-for-communities-of-all-sizes/

Gascon, M., Triguero-Mas, M., Martínez, D., Dadvand, P., Forns, J., Plasència, A., & Nieuwenhuijsen, M. J. (2015). Mental health benefits of long-term exposure to residential green and blue spaces: A systematic review. *International Journal of Environmental Research and Public Health, 12*(4), 4354–4379. https://doi.org/10.3390/ijerph120404354

Ghorbany, S., Hu, M., & Nouri, A. (2025). Commercial buildings decarbonization: Benchmarks and strategies for the United States sustainable commercial construction. *Sustainable Cities and Society, 124*, Article 106324. https://doi.org/10.1016/j.scs.2025.106324

Government of Belo Horizonte. (2025, September 15). *PBH assina protocolo com o Google para projetos de segurança, mobilidade e sustentabilidade na capital*. https://prefeitura.pbh.gov.br/noticias/pbh-assina-protocolo-com-o-google-para-projetos-de-seguranca-mobilidade-e-sustentabilidade

Gregory, A., Spence, E., Beier, P., & Garding, E. (2021). Toward best management practices for ecological corridors. *Land, 10*(2), 140. https://doi.org/10.3390/land10020140

Hautamäki, R., Kulmala, L., Ariluoma, M., & Järvi, L. (2025). How urban green infrastructure contributes to carbon neutrality. *Buildings & Cities, 6*(1), 272–280. https://doi.org/10.5334/bc.586

Hirst, S. (2014). *Zoned in the USA: The origins and implications of American land use regulation*. Cornell University Press.

Houlden, V., Weich, S., Albuquerque, J. P., Jarvis, S., & Rees, K. (2018). The relationship between greenspace and the mental wellbeing of adults: A systematic review. *PLoS ONE, 13*(9), Article e0203000. https://doi.org/10.1371/journal.pone.0203000

Hollis, H., Skropke, C., Smith, H., Harries, H., & Garling, O. (2023). *Social infrastructure: International comparative review.* IFCS & Bennett IFPP. www.bennettinstitute.cam.ac.uk/wp-content/uploads/2023/01/Social_infrastructure_international_comparative_review.pdf

IRP-UNEP. (2018). *The weight of cities. Resource requirements of future urbanization.* www.resourcepanel.org/file/971/download?token=ehOygAQ7

ITF-OECD. (2024). *Urban logistics hubs: Summary and conclusions.* ITF Roundtable Reports. No. 195. OECD. https://www.itf-oecd.org/sites/default/files/docs/urban-logistics-hubs.pdf

Jarrahi, A., Aflaki, A., Khakpour, M., & Esfandiari, M. (2024). Enhancing indoor air quality: Harnessing architectural elements, natural ventilation and passive designs strategies for effective pollution reduction—A comprehensive review. *Science of the Total Environment, 954*, Article 176631. https://doi.org/10.1016/j.scitotenv.2024.176631

Jin, T., Wang, K., Xin, Y., Shi, J., Hong, Y., & Witlox, F. (2024). Is a 15-minute city within reach? *Land Use Policy, 142*, Article 107180. https://doi.org/10.1016/j.landusepol.2024.107180

Kabisch, N., Basu, S., van den Bosch, M., Bratman, G. N., & Masztalerz, O. (2023). Nature-based solutions and mental health. In T. McPhearson, N. Kabisch, & N. Frantzeskaki (Eds.), *Nature-based solutions for cities* (pp. 192–212). Edward Elgar Publishing. https://doi.org/10.4337/9781800376762.00019

Karunakaran, R., Yuen, C. W., Chuah, J. H., Zainol, R., & Goh, B. H. (2023). Correlation analysis and regression of transit ridership by means of public buses—A preliminary data analysis. *IOP Conference Series: Materials Science and Engineering, 1289*, Article 012044. https://doi.org/10.1088/1757-899X/1289/1/012044

Kempe, Y. (2024, September 9). Google pilots heat resilience tool in 14 US cities. *SmartCities Dive.* www.smartcitiesdive.com/news/google-heat-resilience-mapping-tool-cities-trees-cool-roofs/726424/

Keniger, L. E., Gaston, K. J., Irvine, K. N., & Fuller, R. A. (2013). What are the benefits of interacting with nature? *International Journal of Environmental Research and Public Health, 10*(3), 913–935. https://doi.org/10.3390/ijerph10030913

Kerr, G. H., Meyer, M., Goldberg, D. L., Miller, J., & Anenberg, S. C. (2024). Air pollution impacts from warehousing in the United States uncovered with satellite data. *Nature Communications, 15*, 6006. https://doi.org/10.1038/s41467-024-50000-0

Knap, E., Ulak, M. B., Geurs, K. T., Mulders, A., & van der Drift, S. (2023). A composite X-minute city cycling accessibility metric and its role in assessing spatial and socioeconomic inequalities—A case study in Utrecht, the Netherlands. *Journal of Urban Mobility, 3*, Article 100043. https://doi.org/10.1016/j.urbmob.2022.100043

Kumar, P., Druckman, A., Gallagher, J., Gatersleben, B., Allison, S., Eisenman, T. S., Hoang, U., Hama,S., Tiwari, A., Sharma, A., Abhijith, K. V., Adlakha, D., McNabola A., Astell-Burt, T., Feng, X., Skeldon, A. C., de Lusignan, S., & Morawska, L. (2019). The nexus between air pollution, green infrastructure and human health. *Environment International Part A, 133*, 105181. https://doi.org/10.1016/j.envint.2019.105181

Langemeyer, J., & Baró, F. (2021). Nature-based solutions as nodes of green-blue infrastructure networks: A cross-scale, co-creation approach. *Nature-Based Solutions, 1*, Article 100006. https://doi.org/10.1016/j.nbsj.2021.100006

Lee, S., Kim, J., & Cho, K. (2024). Temporal dynamics of public transportation ridership in Seoul before, during, and after COVID-19 from urban resilience perspective. *Scientific Reports, 14*, 8981. https://doi.org/10.1038/s41598-024-59323-w

Li, Y., Zhang, Y., & Yu, X. (2019). Urban weight and its driving forces: A case study of Beijing. *Science of the Total Environment, 658*, 590–601. https://doi.org/10.1016/j.scitoenv.2018.12.022

Li, T., Kent, A., Doson, J., & Sima, Y. (2024). How workable is a 20-minute city in Australia? Re-examining the challenge of change in spatial labour and job relations. *Urban Geography, 46*(5), 1131–1156. https://doi.org/10.1080/02723638.2024.2415795

Lin, Y., Huang, T., Yang, W., Hu, X., & Li, C. (2023). A review on the impact of outdoor environment on indoor thermal environment. *Buildings, 13*(10), 2600. https://doi.org/10.3390/buildings13102600

Liu, D., Kwan, M.-P., & Wang, J. (2024). Developing the 15-minute city: A comprehensive assessment of the status in Hong Kong. *Travel Behaviour and Society, 34*, Article 100666. https://doi.org/10.1016/j.tbs.2023.100666

Liyanage, D. R., Hewage, K., Hussain, S. A., Razi, F., & Sadiq, R. (2024). Climate adaptation of existing buildings: A critical review on planning energy. *Renewable and Sustainable Energy Reviews, 199*, Article 114476. https://doi.org/10.1016/j.rser.2024.114476

Lwasa, S., Seto, K. C., Bai, X., Blanco, H., Gurney, K.R., Kılkış, Ş., Lucon, O., Murakami, J., Pan, J., Sharifi, A., & Yamagata, Y. (2022). Urban systems and other settlements. In *IPCC, 2022: Climate change 2022: Mitigation of climate change. Contribution of working group III to the sixth assessment report of the Intergovernmental Panel on Climate.* Cambridge University Press. https://doi.org/10.1017/9781009157926.010

Mahadevia, D., Delgado Ramos, G. C., Barnes, J., Fitzgerald, J., Kamei, M., & Lanza, K. (2025). Learning from COVID-19 for climate-ready urban transformation. *Cambridge University Press*. https://doi.org/10.1017/9781009527279

Mangone, G., Kurvers, S. R., & Luscuere, P. G. (2014). Constructing thermal comfort: Investigating the effect of vegetation on indoor thermal comfort through a four-season thermal comfort quasi-experiment. *Building and Environment, 81*, 410–426. https://doi.org/10.1016/j.buildenv.2014.07.019

Martínez, S., Sanchez, R., & Yañez-Pagans, P. (2024). Getting a lift: The impact of aerial cable cars in La Paz, Bolivia. *Economía LACEA Journal, 23*(1), 204–229. https://doi.org/10.31389/eco.439

McShane, I., & Coffey, B. (2022). Rethinking community hubs: Community facilities as critical infrastructure. *Current Opinion in Environmental Sustainability, 54*, Article 101149. https://doi.org/10.1016/j.cosust.2022.101149

Megahed, N. A., & Ghoneim, E. M. (2021). Indoor air quality: Rethinking rules of building design strategies in post-pandemic architecture. *Environmental Research, 193*, Article 110471. https://doi.org/10.1016/j.envres.2020.110471

Messier, L. (2025). Complete streets meet fragmented policies: Sidewalks in 30 U.S. cities. *Cities, 157*, 105588. https://doi.org/10.1016/j.cities.2024.105588

Mohammadi-Mavi, H., Fardin, M., & Ranjbari. (2025). Should we blame COVID-19 for the decline in transit ridership, or was it merely a stimulant? *Transportation Research: Interdisciplinary Perspectives, 31*, 101397. https://doi.org/10.1016/j.trip.2025.101397

Monstadt, J., & Saltzman, K. (2025). How data centers have come to matter: Governing the spatial and environmental footprint of digital gateway to Europe. *International Journal of Urban and Regional Research, i*(4), 757–778. https://doi.org/10.1111/1468-2427.13316

Moreno, C., Allam, Z., Chabaud, D., Gall, C., & Pratlong, F. (2021). Introducing the "15-minute city": Sustainability, resilience and place identity in future post-pandemic cities. *Smart Cities, 4*(1), 93–111. https://doi.org/10.3390/smartcities4010006

Nelson, A. C., & Hibberd, R. (2023). Complete streets as a redevelopment strategy. *Cityscape: A Journal of Policy Development and Research, 26*(2), 311–382. https://www.huduser.gov/portal/periodicals/cityscape/vol26num2/ch17.pdf

Neves, A., Godina, R., Azevedo, S. G., Pimentel, C., & Matias, J. C. O. (2019). The potential of industrial symbiosis: Case analysis and main drivers and barriers to its implementation. *Sustainability, 11*(24), 7095. https://doi.org/10.3390/su11247095

Nugroho, N. Y., Triyadi, S., & Wonorahardjo, S. (2022). Effect of high-rise buildings on the surrounding thermal environment. *Building and Environment, Part A, 207*, Article 108393. https://doi.org/10.1016/j.buildenv.2021.108393

Organisation for Economic Cooperation and Development (OECD). (2021). OECD regional outlook 2021: Addressing COVID-19 and moving to net zero greenhouse gas emissions. *OECD Publishing*. https://doi.org/10.1787/17017efe-en

Organisation for Economic Cooperation and Development (OECD). (2024). *Sustainable accessibility for all.* OECD. www.oecd.org/en/publications/sustainable-accessibility-for-all_5c91857c-en.html

Ouyang, Z., Sciusco, P., Jiao, T., Feron, S., Lei, C., Li, F., John, R., Fan, P., Li, X., Williams, C. A., Chen, G., Wang, C., & Chen, J. (2022). Albedo changes caused by future urbanization contribute to global warming. *Nature Communications, 13*, 3800. https://doi.org/10.1038/s41467-022-31558-z

Pabilonia, S. W., & Redmond, J. J. (2024, October). The rise in remote work since the pandemic and its impact on productivity. *Beyond the Numbers: Productivity, 13*(8), U.S. Bureau of Labor Statistics. www.bls.gov/opub/btn/volume-13/remote-work-productivity.htm

Paris Government. (2022). *Paris ville du Quart d'heure, ou le pari de la proximité.* https://www.paris.fr/dossiers/paris-ville-du-quart-d-heure-ou-le-pari-de-la-proximite-37

PNUMA – Programa de las Naciones Unidas para el Medio Ambiente. (2021). *El Peso de las Ciudades en América Latina y el Caribe: Requerimientos futuros de recursos y potenciales rutas de actuación.* https://wedocs.unep.org/bitstream/handle/20.500.11822/37655/PCLAC.pdf

Prieto-Rodriguez, J., Azuela-Flores, J. I., Groft, D., Perez-Villadoniga, M. J., & Salas, R. (2025). Assessing the impact of the new Mexico cable car on air pollution. *Journal of Transport Geography, 122*, Article 104052. https://doi.org/10.1016/j.jtrangeo.2024.104052

Rao, F., Kong, Y., Ng, K. H., Xie, Q., & Zhu, Y. (2024). Unravelling the spatial arrangement of the 15-minute city: A comparative study of Shanghai, Melbourne and Portland. *Planning Theory & Practice, 25*(2), 184–206. https://doi.org/10.1080/14649357.2024.2350948

Raven, J., Leone, M. F., Bhaduri, S., Braneon, C., Corbett, D., Driskell, D, icker, U., Fernandez, J. E., Jing, G., Hürlimann, A., Judah, I., & Rosenzweig, C. (2025a). *Planning, urban design, and architecture for climate action.* Cambridge University Press.

Raven, J., Leone, M. F., & Nocerino, G. (2025b). The urban design climate workshop: Integrating mitigation and adaptation strategies to confront urban climate change. In B. He, Y. Wang, & A. Cheshmehzangi (Eds.), *Urban climate and urban design* (pp. 167–179). Springer Nature Singapore.

Renner, A. T., Plank, L., & Getzner, M. (2024). *Handbook of social infrastructure. Conceptual and empirical research perspectives*. Edward Elgar Publishing.

Reynolds, H. L., Brandt, L., Fischer, B. C., Hardiman, B. S., Moxley, D. J., Sandweiss, E., Speer, J. H, & Fei, S. (2020). Implications of climate change for managing urban green infrastructure: an Indiana, US case study. *Climatic Change, 163*, 1967–1984. https://doi.org/10.1007/s10584-019-02617-0

Ruini, A., Sporchia, F., Niccoluci, V., Pulselli, F. M., & Bastianoni, S. (2025). Rethinking environmental benefit allocation in industrial symbiosis. *Science of the Total Environment, 992*, Article 179932. https://doi.org/10.1016/j.scitotenv.2025.179932

Shanghai Municipal Government. (2024). *Shanghai 15-minute community life circle action plan*. https://ghzyj.sh.gov.cn/cmsres/3a/3a52297af92f43a29899e3d61e4c5cc2/36def90629a5427faec623ed7f2b4dbb.pdf

Shi, Y., Sorrell, S., & Foxon, T. (2023). The impact of teleworking on domestic energy use and carbon emissions: An assessment for England. *Energy and Buildings, 287*, Article 112996. https://doi.org/10.1016/j.enbuild.2023.112996

Shoina, M., Voukkali, I., Anagnostopoulos, A., Papamichael, I., Stylianou, M., & Zorpas, A. A. (2024). The 15-minute city concept: The case study within a neighbourhood of Thessaloniki. *Waste Management & Research, 42*(8), 694–710. https://doi.org/10.1177/0734242X241259926

Simon, S., & O'Brien, W. (2023). Pilot study to measure the energy and carbon impacts of teleworking. *Buildings & Cities, 4*(1), 174–192. https://doi.org/10.5334/bc.271

Schmitt, P., & Wiechmann, T. (2018). Unpacking spatial planning as the governance of place. *Planning Review, 54*(4), 21–33. https://doi.org/10.1080/02513625.2018.1562795

Solecki, W., Delgado Ramos, G. C., Roberts, D., Rosenzweig, C., & Walsh, B. (2021). Accelerating climate research and action in cities through advanced science–policy–practice partnerships. *Urban Sustainability, 1*, 3. https://doi.org/10.1038/s42949-021-00015-z

Steer & Fourth Economy. (2024). Efficiency and emissions impact of last mile online delivery in the U.S. https://progresschamber.org/wp-content/uploads/2024/06/Chamber-of-Progress-Efficiency-and-Emissions-Impact-of-Last-Mile-Online-Delivery-in-the-US.pdf

United Nations, Department of Economic and Social Affairs, Population Division. (2019). *World urbanization prospects: The 2018 revision* (ST/ESA/SER.A/420).

United Nations Environment Programme (UNEP) and UN-Habitat. (2021). *Global environment for cities—GEO for cities: Towards green and just cities*. UNEP. https://wedocs.unep.org/bitstream/handle/20.500.11822/37413/GEOcities.pdf

UN-Habitat – United Nations Human Settlements Programme. (2020). *More than four walls and a roof: Adequate housing, a human right that can unlock many others.* https://unhabitat.org/more-than-four-walls-and-a-roof-adequate-housing-a-human-right-that-can-unlock-many-others

USDN – Urban Sustainability Directors Network. (2019). *Guide to developing resilience hubs. Urban Sustainability Directors Network.* http://resilience-hub.org/wp-content/uploads/2019/10/USDN_ResilienceHubsGuidance-1.pdf

Van Nes, A. (2021). Spatial configurations and walkability potentials: Measuring urban compactness with space syntax. *Sustainability, 13*(11), 5785. https://doi.org/10.3390/su13115785

Van Renterghem, T. (2019). Towards explaining the positive effect of vegetation on the perception of environmental noise. *Urban Forestry & Urban Greening, 40*, 133–144. https://doi.org/10.1016/j.ufug.2018.03.007

Viu-Roig, M., & Alvarez-Palau, E. J. (2020). The impact of E-Commerce-related last-mile logistics on cities: A systematic literature review. *Sustainability, 12*(16), 6492. https://doi.org/10.3390/su12166492

Vranayova, Z., Kaposztasova, D., & Lis, A. (2024). Cognitive and enviromental benefits of blue-green infrastructure and green skills training. *Selected Scientific Papers – Journal of Civil Engineering, 19*(1), 1–8. https://doi.org/10.2478/sspjce-2024-0015

Wang, X., Valenzuela, E. A., You, C., Nguyen, V., Sun, L., & Miranda-Moreno, L. F. (2025). The future of urban cycling: A predictive framework for climate change. *Transportation Research Part D: Transport and Environment, 143*, Article 104722. https://doi.org/10.1016/j.trd.2025.104722

WHO – World Health Organization. (2025, March 19). *Urban health.* https://www.who.int/news-room/fact-sheets/detail/urban-health

Williams, J. (2022). Circular cities: Planning for circular development in European cities. *European Planning Studies, 31*(1), 14–35. https://doi.org/10.1080/09654313.2022.2060707

Wong, T. L. X., Lim, E. L., Hasan, M. R., Sougui, O. O., Milad, A., & Qu, X. (2024). Effectiveness of heat-reflective asphalt pavements in mitigating urban heat islands: A systematic literature review. *Journal of Road Engineering, 4*(4), 399–420. https://doi.org/10.1016/j.jreng.2024.04.008

Zannat, E. U., Dedekorkut, A., & Morgan, E. A. (2024). A review of nature-based infrastructures and their effectiveness for urban flood risk mitigation. *Wires Climate Change, 15*(5), Article e889. https://doi.org/10.1002/wcc.889

Zarei, M., & Shahab, S. (2024). Nature-based solutions in urban green infrastructure: A systemic review of success factors and implementation challenges. *Land, 14*(4), 818. https://doi.org/10.3390/land14040818

CHAPTER 11

Systems Approaches to City Research and Innovation: Finance, Governance, and Multi-level Partnerships: Action, Innovation, and Research

Andy Deacon, Eugenie L. Birch, Joanna Masic, Asma Jhina, Johanna Granados Alcala, Marcus Arcanjo, Pourya Salehi, Mauricio Rodas, Aikaterini Karapantsiou, Agathe Cavicchioli, Natalia Salazar, Priscilla Negreiros, Asif Nawaz Shah, and Anna Koch

11.1 Introduction

On September 5, 2024, *Euronews*'s lead article screamed, "The southern state of Rio Grande do Sul [Brazil] has been virtually cut off from the rest of the country as flooding washed away roads and bridges and inundated airports" (Frost, 2024). The news agency was reporting on the four days

A. Deacon (✉) · A. Jhina · J. Granados Alcala · A. Cavicchioli · A. N. Shah · A. Koch
Global Covenant of Mayors for Climate and Energy, Brussels, Belgium
e-mail: adeacon@globalcovenantofmayors.org

C. Oke et al. (eds.), *A Global Research and Action Agenda on Cities and Climate Change Science*,
https://doi.org/10.1007/978-981-95-6517-7_11

of heavy rains that had overwhelmed the state's capital, Porto Alegre, with a population of 1.5 million, as well as more than 400 surrounding cities and towns. Subsequent accounts detailed how all public transport, from air to bus and metros, remained closed for months. Power

A. Jhina
e-mail: ajhina@globalcovenantofmayors.org

J. Granados Alcala
e-mail: jgranadosalcala@gcomprojectsupport.org

A. Cavicchioli
e-mail: acavicchioli@globalcovenantofmayors.org

A. N. Shah
e-mail: ashah@globalcovenantofmayors.org

A. Koch
e-mail: akoch@globalcovenantofmayors.org

E. L. Birch · M. Rodas
University of Pennsylvania, Philadelphia, PA, USA
e-mail: elbirch@design.upenn.edu

M. Rodas
e-mail: rodas@upenn.edu

J. Masic
World Bank, Washington, DC, USA
e-mail: jmasic@worldbank.org

M. Arcanjo
C40 Cities, London, UK
e-mail: marcanjo@c40.org

P. Salehi · N. Salazar
ICLEI - Local Governments for Sustainability, Bonn, Germany
e-mail: pourya.salehi@iclei.org

N. Salazar
e-mail: natalia.salazar@iclei.org

A. Karapantsiou
Deutsche Gesellschaft für Internationale Zusammenarbeit (GIZ), Bonn, Germany
e-mail: aikaterini.karapantsiou@giz.de

was gone, including telecommunications. Neighborhoods were swamped; schools, shops, and offices were shuttered. Shortages in food, medical aid, and social services rapidly emerged. Waterborne diseases increased (Michaelides et al., 2025; Michelon et al., 2025). This disruption affected the state's interconnected regional infrastructure, leading to $15 billion in economic losses, 183 fatalities, and the displacement of over 600,000 people (Simoes-Sosa et al., 2025).

More than a year would pass before the area reached some semblance of normality. Ultimately, the disaster would result in a nearly 2% loss to the state's GDP, later reflected in the overall Brazilian GDP. Simply stated, this case shows that in cities, a breakdown in one system is rapidly followed by failures in others, resulting in enormous economic, environmental, and social costs (Dodman et al., 2022). In the aftermath, observers noted that had the national, state, and local governments collaborated on developing and implementing an integrated climate-resilient infrastructure action plan, they could have largely averted the cascading and compounding effects of this disaster (Molina, 2025).

This story has been repeated in cities around the world. Notable examples include the October 2024 reports of intense rains associated with the high-altitude isolated depression phenomenon, affecting Spain's Valencia region, as well as the June 2025 flooding in Bangladesh that displaced nearly two million urban residents. Reports about the effects of "climate whiplash" (flooding and drought cycles) occurring in cities like Cairo, Jakarta, Beijing, and Bogotá also remain prominent (Michaelides et al., 2025; Regan, 2024). These issues go well beyond large cities. As of 2024, 2214 cities and local governments had reported flooding or sea-level rise as one of their top hazards, while 2122 had cited extreme rainfall as having estimated economic costs amounting to €15 million per hour that, without action, could lead to cumulative losses of 16–22% of global GDP by 2100 (Global Covenant of Mayors, 2024). As the United Nations Framework Convention on Climate Change's (UNFCCC) first Global Stocktake revealed, current methods have proved insufficient to achieve the Paris Agreement goals (UNFCCC, 2024). The world is on course to overshoot the 1.5 °C target.

P. Negreiros
Cities Climate Finance Leadership Alliance, London, UK
e-mail: priscilla.negreiros@cpiglobal.org

The narrative above underlines the importance of looking at cities to address immediate climate-induced disasters and deal with their long-term threats. Thousands of cities have expressed the willingness to undertake the necessary mitigation programs. 75% of their climate-action plans and climate pledges have included commitments to deeper levels of emissions reductions than those of their respective national governments; over 50% have committed to deliver the reductions faster than their respective nations. If achieved, these city pledges could reduce some 40% of global emissions (Global Covenant of Mayors, 2024). Further, in its recent setting of the global goal on adaptation (GGA), the UNFCCC recognized the critical role of subnational areas (COP 27 President, 2023a, 2023b). To date, more than 13,000 cities in 200 regions have included adaptation in their climate plans (Mylnar & Belmonte, 2024).

A focus on cities suggests paying attention to their urban systems, as briefly described in the introduction to Section 3 in this book. It calls for recognizing that most countries—whether unitary or federal in form—have designated competencies for subnational units that allow for multi-level collaborations, integrated stakeholder processes, and scaled-up financing arrangements, all executed if political will exists. Multiple stakeholders, along with some national governments, are developing these ideas and practices and seeking ways to finance them. They have circulated their ideas at the Innovate4Cities conferences and beyond. The results are contributing new knowledge and insights into the Global Research and Action Agenda on Cities and Climate Change Science (GRAA) (Irvin et al., 2024) (see Chapter 3).

In contributing to the GRAA, this chapter's four parts examine urban-systems approaches in relation to finance, governance, and multi-level partnerships. Section 11.2 illuminates the current state and future needs for city-focused infrastructure investment with assessments by region and level of development. Section 11.3 discusses the rationale, progress, and examples of efforts to integrate climate-friendly urban systems finance throughout all spheres of government. Section 11.4 offers examples of matters discussed in Sects. 11.2 and 11.3. Section 11.5 provides suggestions for further exploration in accordance with the findings of this and the previous two chapters and as guided by the updated GRAA.

11.2 Financing Urban Mitigation and Adaptation

Globally, climate-related disasters, including hurricanes, floods, and wildfires, caused an estimated $299 billion in capital asset losses in 2022 alone, resulting in significant effects on national and local GDPs (AON, 2023). This monetary figure, coupled with costs of the floods in Rio Grande do Sul discussed above, prompts the question of whether the cost of inaction can exceed investments in resilient infrastructure. The answer is yes. For example, 2023 estimates for upgrading Porto Alegre's flood-control system amounted to a mere $74 million; a year later the recorded flood damages reached some $300 million (Maciel, 2024; Rodrigues, 2024). Finally, addressing global-warming-induced infrastructure damages goes beyond remedying current losses to contribute in the long run to a city's economic development while offering multiple social and environmental benefits (Heubaum et al., 2022; Tanner et al., 2019).

11.2.1 Urban Climate Finance: Current State

Global urban climate-finance flows in 2022 amounted to $831 billion, a sum dominated by mitigation (CCFLA, 2024). Notably, private finance provided 49% of this total, attributed to purchases of energy-efficient household appliances and electric vehicles. Spending by the public sector constituted 22% of the total (CCFLA, 2024) (Fig. 11.1).

The distribution of climate finance is deeply uneven across regions and sectors. Most flows (81%, or $675 billion) occurred in the Global North and China with 11% ($88 billion) in the Global South. Sub-Saharan Africa (1% or $5 billion) was critically underfunded (CCFLA, 2024). In terms of sectors, transportation captured nearly 50% of the total, and building energy 28%.

11.2.2 Urban Climate Finance: Future Need

The World Bank estimates the future capital costs of resilient and low-carbon urban systems in low-and middle-income countries to be between $256 and $821 billion per year through 2050. This number represents 0.8–2.6% of these countries' combined GDPs. In these projections, the East Asia Pacific region has the highest absolute costs, while the Sub-Saharan African region has the greatest share of GDP (CCFLA, 2024). These estimates assume that the public sector alone cannot provide

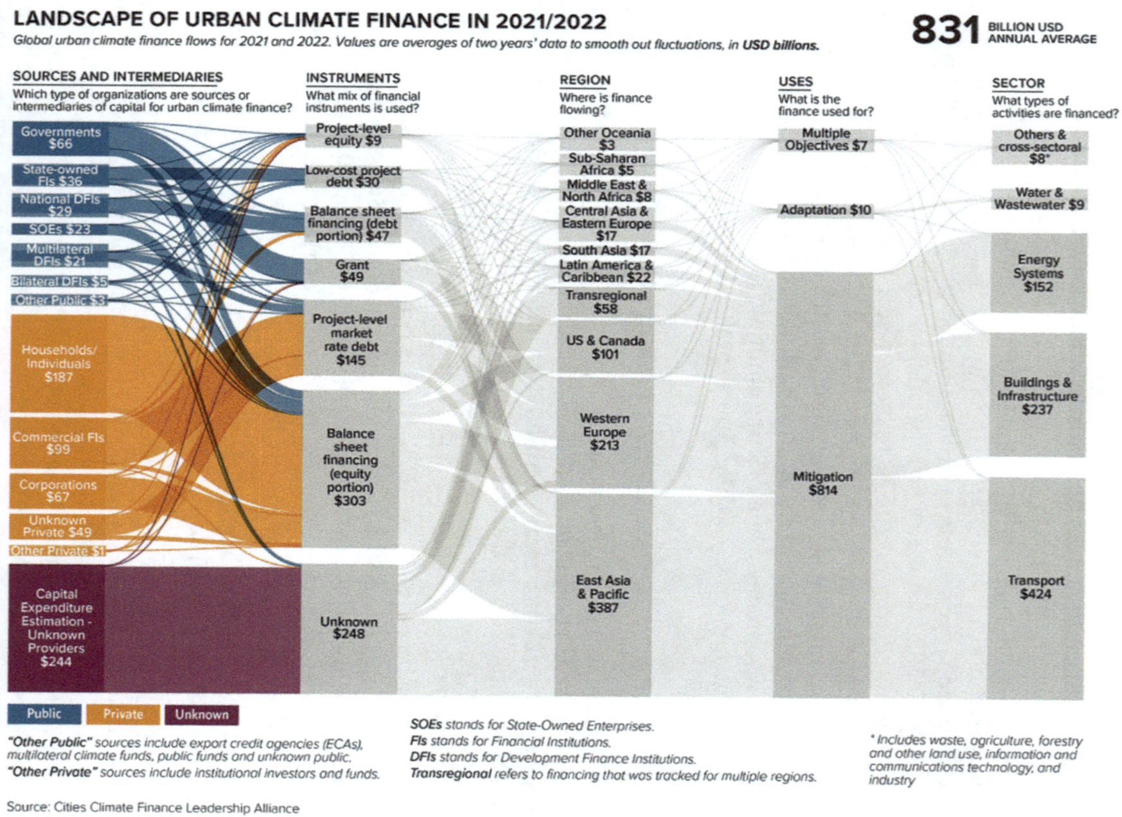

Fig. 11.1 In 2022, private-sector spending was higher than that of the public sector due to household investments in EVs and energy-saving appliances (*Source* CCFLA, 2024, p. 10, used with permission)

sufficient capital investments, suggesting that that private-sector and philanthropic contributions will be necessary. Not to be forgotten are capital projects' annual operating and maintenance costs, a figure that ranges between $525 billion and $548 billion (Deuskar et al., 2025). The estimates further assume that funding associated with operations/ maintenance will come from several sources, depending on the city, including own-source revenues, intergovernmental transfers, cost savings, and monetizing the benefits of investments.

11.2.3 *Addressing the Gap in Urban Climate Financial Flows*

It is increasingly clear that the world has failed to meet the 1.5 degree temperature target. The reality that cities are responsible for 70% of global greenhouse gas emissions, coupled with the shockingly low financial flows to urban climate change efforts, has underscored a universal need for new urban climate solutions. While increasing investment in cities is essential, a surge in funding represents only a partial answer. Current understanding of how cities operate and grow suggests the importance of city-scale strategies to capture cascading and multiplicative effects embodied in an urban-systems approach (Lwasa et al., 2022; World Economic Forum, 2022).

11.2.4 *An Urban-Systems Approach to Urban Climate Finance*

In general, a systems approach leverages "the interconnected and interdependent relationships—including reinforcing and balancing feedback...[and] recognizes the interdependence and differing reaches among systems. Interventions in one sector can generate effects across others" (World Economic Forum, 2022).

While it is admittedly hard to execute for a city, an urban-systems approach, often called climate-resilient development (Schipper et al., 2022), entails the horizontal and vertical integration of the city's various sectors within an overall plan. It requires fostering cooperation and alignment on the planning, execution, and operation of climate-sensitive policies and associated programs within and among all levels of government and among relevant stakeholders. It involves several sectors as already explored in Chapter 10: transportation, building energy, solid-waste management, water and wastewater, flood protection (coastal, riverine, and pluvial), heat resilience, and digital infrastructure, see Fig. 11.2

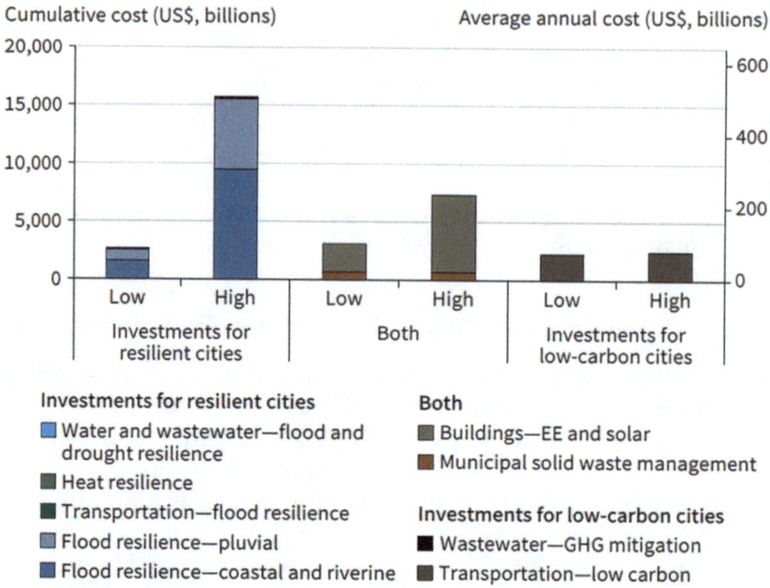

Fig. 11.2 By income group, upper-middle-income countries have the highest costs in absolute terms; low-income countries have the highest costs in terms of share of GDP (*Source* Deuskar et al., 2025, p. 8, used with permission)

(Deuskar et al., 2025; World Economic Forum, 2022). This form of investment frequently addresses the dual challenges of mitigation and adaptation.

Building climate-resilient infrastructure tends to be capital-intensive; moreover, once built, it is long-lasting, requiring continued investment in operations and maintenance. For example, building a single wastewater-treatment plant serving 10,000 people costs ~ $12 million/unit, with a 40- to 50-year lifespan (US EPA, 2025; Mahendra et al., 2021 Expanding this example to a holistic infrastructural system reveals the challenges of the approach.

11.2.5 Critical Urban Systems

Transport and buildings, the largest sources of greenhouse gases, are the first line of attack. Further, as inter-related systems, they have a far greater effect on greenhouse-gas reduction and adaptation if treated together than if treated independently. In addition, these treatments involve other systems; energy, water, sanitation, solid-waste disposal, and open space.

One of the most effective greenhouse-gas-reducing techniques within an urban-systems approach is to substitute multi-modal transportation (renewable-energy-powered mass transit, bicycle, and pedestrian) for single mode (fossil-fueled cars). However, to be economically and functionally successful, this policy needs to include implementation of a compact and transboundary land-use pattern that incorporates a series of urban systems: high-density mixed-use buildings (to provide demand for such greenhouse-gas-reducing elements as mass transit, district heating, and walkable commerce or jobs), open space (to accommodate bicycles and pedestrian traffic and facilitate CO_2 capture), water, sanitation, solid waste (using a range of climate-friendly solutions, nature-based solutions including green infrastructure, and waste-to-energy programs), and food and water supply. Energy transition is at the heart of the complete effort. Beyond mitigation, planning the location and design of built environment and infrastructure systems in compact developments leads to adaptation. Financing this approach supports a range of public and private investments that could be laid out in a portfolio (RCN, 2025). It also involves international cooperation programmes that promote solution-oriented investments, such as the Euroclima+, a European Union cooperation initiative focused on promoting sustainable, low carbon transport for transforming urban mobility.

UK's International Climate Finance (ICF) Programme on sustainable infrastructure as well as its Climate Fund Accelerator are other examples of international cooperation with significant impacts at the local level, the latter reporting 410 million dollars in investments mobilized between 2020 and 2024 for 238 projects in ten countries. Another example is Germany's International Climate Initiative (IKI), recently more focused on energy transition, sustainable mobility, and nature-based solutions. Its Integrated Urban Climate Action for Low-Carbon & Resilient Cities project serves as an example in the Asia–Pacific region while, in the Latin American region, Germany's KfW and BMZ, in collaboration with ICLEI, are also supporting cities to access more affordable insurance coverage against climate change impacts through the Urban Infrastructure Insurance Facility.

11.3 Basic Requirements for Planning, Executing and Operating an Urban-Systems Approach

An urban-systems approach necessitates a development framework; one secured through strong urban planning that aligns climate objectives with the city's and upper-level governments' broader economic and social priorities, including gender, equity, cultural and heritage considerations (see Chapter 9). A plan coordinates urban systems comprehensively. It calls for understanding the relationship between greenhouse-gas production and spatial development, creating multi-sector solutions for mitigation and adaptation. It includes stakeholders in productive dialogues in prioritizing interventions.

Managing growth is a fundamental challenge to the systems approach in rapidly urbanizing places in the Global South. Most places are experiencing higher rates of land consumption than rates of population growth, often as a proliferation of informal settlements. With this phenomenon occurring along coastlines or hillsides or impinging on watersheds and agriculture, sprawling, energy-consuming land-use patterns are the result (Angel et al., 2016; Behnisch et al., 2022; Mahendra & Setom, 2019). Combating this situation calls for such multi-level government actions as creating national urban-policy directives, climate-sensitive infrastructure funding, and supporting local urban planning that encompasses compact-development policies and biodiversity conservation (see Chapters 10 and 9, respectively).

Unlocking urban-systems-focused climate finance also requires supportive national law. Cities need direct access to such capital-market instruments as municipal bonds and loans, and guarantees, including the ability to borrow without a sovereign guarantee and permission to engage in public–private partnerships and special purpose vehicles. It should be noted that any local government debt carrying a sovereign guarantee, including debt incurred by cities, becomes a national contingent liability included in a country's overall debt that the IMF reviews in its national debt-sustainability assessment, a measure that determines a nation's access to IMF financing programs. This condition further underlines the necessity of having smooth multi-level governance as a critical element in urban climate finance. Thus, while local governments need to mobilize resources, they must do so within the bounds permitted by their nations.

A city needs to be creditworthy to access capital markets. This means having an investment-grade rating either from such international agencies as Moody's, S&P, and Fitch, or from within-country agencies. Examples of domestic ratings companies are Global Credit Ratings (GCR South Africa), Agusto & Co. (Nigeria), CRISIL (India), HR Ratings (Mexico), and Austin Ratings (Brazil). Ratings build investor confidence and trust by attesting that a city can meet its financial obligations; they also determine the terms (interest rate, length, and amount) of loans. The higher the rating, the better the terms.

As urban climate finance centers on long-term capital-intensive infrastructure investments, access and terms are extremely important, yet few cities have ratings. A BBB or BBB- rating from an international agency opens global capital markets and the flow of foreign direct investment. A rating from a domestic agency facilitates borrowing on the local market, where the available amounts may be limited. Yet, fewer than 4% of the 500 largest cities in the Global South have globally recognized credit ratings and fewer than 20% are creditworthy in their local contexts (World Bank Foundation, 2015).

Achieving creditworthiness is extremely challenging. Based on sound municipal fundamentals, it hinges on consistent and reliable revenue generation combined with disciplined expenditure management, a situation that requires implementing clear, publicly accessible budgeting and financial reporting (World Bank, 2025). Maximizing own-source revenues (e.g., efficient tax collection, utility billing), having strong governance and institutional capacity (e.g., skilled financial-management teams and robust internal controls), and engaging in long-term financial planning (e.g., anticipating risk and future needs) result in a bankable credit rating (White & Wahba, 2019). Importantly, differences in city capacities and available resources will significantly affect any ability to undertake the actions listed above.

11.3.1 Scaling Up Urban Climate Finance

Scaling up urban climate finance not only requires developing (and implementing) urban-systems approaches but also linking local, regional, national, and even international frameworks to enable a move away from isolated, project-by-project financing toward programmatic, transformative interventions (Boex et al., 2021; World Economic Forum, 2022). This conversely requires greater inclusion of cities in national and international climate-finance dialogues relating to the reformation of the global financial architecture, and the recognition of cities' central role in climate transitions.

One scaling-up technique currently being discussed among multi-level development banks and others is the use of country platforms: nationally led mechanisms to foster collaboration among partners sharing a common vision and priorities (MDB Action Group, 2024). South Africa, Egypt, Indonesia, and Bangladesh have developed them to address energy transition involving national agencies (Gilmour et al., 2024). City advocates, including the City Climate Finance Leadership Alliance and the SDSN Global Commission on Urban SDG Finance, have called for including urban sectors as key priorities within these actions and having city voices represented in their governance structures to ensure coverage of urban investments (CCFLA & SDSN Global Commission, 2025). Such platforms need to recognize cities' and towns' different needs, abilities, and mandates regarding infrastructure provision, even if they are in the same country.

Another technique to scale up urban climate finance is aggregating projects into larger, investment-ready portfolios. Bundling multiple projects can lower transaction costs, reduce investment risks, take advantage of city procurement purchasing power, and attract capital to projects with a higher ticket size. Programmatic financing frameworks and standardization tools—such as protocols for monitoring, reporting, and verification, and risk-mitigation instruments—further enable this scaling by building bankable project pipelines. Such bundling of projects could happen at an individual city scale or potentially across multiple cities in a country. Creating such coordination mechanisms as high-level task forces, investment facilities, financial intermediaries, or funds can assist in facilitating resource flows into aggregated portfolios within and across levels of government (Pinko et al., 2022).

11.4 Enabling Effective Multi-level Governance

Increasingly, cities are critical in helping translate Long-Term Low-Emission Development Strategies, Nationally Determined Contributions, and National Adaptation Plans (NAPs) agreed to in the Paris Agreement into concrete action delivered at the pace and scale needed (Revi et al., 2022). But cities cannot act alone: they need to be incorporated into a national model set up to distribute and coordinate roles, responsibilities, and instruments among levels of governments to deliver their share of national efforts (GIZ, 2018). Such multi-level governance arrangements improve policy coherence, maximize synergies, navigate trade-offs, promote equity, social justice, and resilience (Coalition for Urban Transitions, 2019). To be effective, stakeholder coordination and feedback mechanisms in multi-level governance systems are key (Global Covenant of Mayors, 2024).

The *Paris Agreement* did encourage multi-level governance and successive declarations (the Glasgow Climate Pact [2021], the Sharm el Sheikh Implementation Plan [2022], the UAE Consensus [2023], and the Baku Climate Unity Pact [2024]) continued the support. Nations interpret these statements in their NDCs. A recent assessment of urban content in the 2025 NDCs showed that 78% of the 194 submissions had strong urban content, increasing again from 2023 and 2017, however only 13% (26) identified the need for urban climate finance (Ranalder et al., 2025; UN Habitat et al., 2024).

In working to translate multi-level governance rhetoric into practice, COP presidents, UN agencies, and city networks have hosted parallel ministerial meetings on urbanization and climate at COP meetings. The Egyptian COP 27 president convened the first. For the second, the UAE COP 28 president managed to get a Local Climate Action Summit included in the official convenings (COP 28 President, 2023). However, successive presidents of COP 29 and 30 have reverted to parallel scheduling.

Despite their positioning, these sessions have yielded programs to draw attention and enhance multi-level governance and partnerships. COP 27 produced "Sustainable Urbanisation for the Next Generation (SURGe)," developed in collaboration with UN-Habitat and ICLEI, endorsed by 180+ global partners to assist urban-friendly national governments in integrating subnational programs (COP 27 President, 2023b). COP 28 launched CHAMP (Coalition for High Ambition Multi-level Partnership for climate action) in partnership with Bloomberg Philanthropies. By 2025, 77 national governments and the EU joined, pledging to; (1) consult and collaborate with subnational governments over climate action with a first major milestone to include contributions to Nationally Determined Contributions 3.0, due in at COP30 2025, and (2) include investment needs and priorities from subnational governments in national investment priorities and help unlock climate finance, including public and private funding, to support subnational governments on delivering ambitious climate action (CHAMP, 2025). Azerbaijan's COP 29 added the Multisectoral Action Pathways (MAP) for Resilient and Healthy Cities, endorsed by 160 stakeholders—nations, cities, UN agencies, civil society organizations—to further enhance multi-level participation in urban climate solutions (COP 29 President, 2024).

Achieving multi-level governance requires capacity-building. Several efforts to provide assistance have emerged. They include the Global Covenant of Mayors' *Multilevel Climate Action Playbook: Policies and Enablers for All Levels of Government* (2022), C40 Cities' *NDC Ambition Handbook: Learning from City Success to Raise National Ambition* (2022), and UN-Habitat's *Seizing the Urban Opportunity: A 3-step Guide for Strengthening Nationally Determined Contributions 3.0* (2025). In 2025, ICLEI launched the COP Town Hall initiative to support community-led convenings aimed at translating global goals to a local context. The Marrakech Partnership for global climate action (MPGCA) dating from COP22 is the foundation for many of these activities.

In enabling multi-level governance, international cooperation has also played an important role. Since 2018, UK PACT—partnering for accelerated Climate Transitions has supported 18 countries of the Global South to build capacities to achieve their National Determined Contributions, finance projects (89 million pounds of committed funding), and advance and even embed green finance within the national financial regulator as in the case of Mexico. Similarly, GIZ's funding for improving the regulatory and financial framework to accelerate public and private climate finance through Mexico's FINACC programme, can be mentioned. The Government of Mexico has estimated that achieving its NDCs would require mobilizing approximately US$126 billion during 2014–2030 (Instituto Nacional de Ecología y Cambio Climático, 2018).

11.5 Selected Case Studies

The following case studies illustrate themes discussed in this chapter: city-focused climate-finance instruments, technical assistance to help cities to develop climate-sensitive plans and projects, compact development, and multi-level governance programs.

11.5.1 Green and Sustainability Bonds

In 2024, the Green and Sustainability (GSS) bond market captured more than $1 trillion (Climate Bond Initiative, 2025). While GSS bonds represent a small fraction of the $130 trillion global bond market, they have experienced continued growth since 2010, with more than $6.2 trillion in issuances since 2010 (Climate Bond Initiative, 2025). Issuers of GSS bonds range from national to local governments and from multilateral development banks to corporations and commercial banks.

The ASEAN countries have a higher share (58%) of government-issued GSS compared to the global share (39%). Malaysia, Singapore, and Indonesia have been active in this area. Each has a distinct policy and regulatory framework. While Malaysia promotes green-bond issuance through its Sustainable and Responsible Investment (SRI sukuk) framework with tax exemptions, Singapore reduces the cost of issuance through the Sustainable Bond Grant Scheme and encourages global participation by adopting international standards. Indonesia has issued a sustainable finance framework, as well as green bonds and a green sukuk framework.

At the local level, the first municipal green bonds were issued in 2013 for Gothenburg, Sweden and the Commonwealth of Massachusetts in the US. Of the Global South countries, Mexico City issued the first $50 million bond, to support climate-friendly transport improvements; Manila followed with a $500 million sustainability bond from Manila Water. Subsequent issuances have reinforced the viability and scalability of this financing mechanism, evidenced by Mexico City's follow-up joint issuance of $100 million bond and Reykjavík's cumulative green-bond volume reaching $130 million prior to 2022 (Climate Bond Initiative, 2023). These cases demonstrate increased investor confidence and institutional capacity in both emerging and developed markets. The potential for scaling remains substantial, indicating that municipal green bonds could play an increasingly central role in financing sustainable and climate-resilient urban infrastructure.

11.5.2 Tools to Aid in Urban Systems Finance

Several tools and platforms exist to support cities in adopting a systems-based approach to finance. For instance, ICLEI's Transformative Actions Program (TAP), launched in 2015, aims at enhancing access to finance for the implementation of ambitious, locally led sustainability projects.

To date, TAP has supported more than 160 projects across more than 50 countries, focusing on areas such as clean energy, sustainable mobility, nature-based solutions, and climate resilience. Another example is the CDP-ICLEI Unified Reporting System, a global platform that enables cities to report climate data in a standardized format, supporting transparency, benchmarking, and access to climate finance. The World Bank Group's APEX Green City tool and City Resilience Resilience Program support cities in preparing green action plans and help prioritize urban resilience investments, respectively. The European Bank for Reconstruction and Development (EBRD) Green Cities program offers several programs, including technical assistance to develop Green City Action Plans, project identification, feasibility, financial modeling, and implementation.

11.5.3 The City Climate Finance Gap Fund: Accra, Ghana, Cape Town, South Africa, and Central Bosnia Canton, Bosnia

Created in 2020 with an initial funding of €100 million and replenished in 2025 with another €50 million, the Gap fund is focused on supporting technical assistance for early-stage project development. Administered by the World Bank and the European Investment Bank, it has supported projects in more than 300 cities. Work in Accra, Ghana, and Cape Town are examples.

Currently, Accra relies heavily on open landfills without proper control and sanitary conditions, leading to environmental degradation, greenhouse gas emissions, and inefficient resource usage. Funded by a grant from the Gap fund, the Accra Metropolitan Assembly began translating a waste-to-biogas concept into a viable, investment-ready proposal via a prefeasibility study that addresses feedstock analysis, biogas technologies, financial modeling, environmental and social impacts, and capacity-building. Also addressed are gender-responsive planning and the inclusion of informal waste workers, aligning with Accra's Climate Action Plan and Ghana's NDCs.

In Cape Town, the Gap fund supported the city's Technical Partnership Unit in developing a flood-control initiative. The Gap Fund is now providing technical support to the city for the development of catchment storm-water master plans to improve the understanding of flood risk and develop a portfolio of flood- alleviation measures.

The majority of Central Bosnia Canton's 250,000 population in 12 municipalities are highly dependent on wood and coal for heating, resulting in high levels of air pollution, CO_2 emissions, and the attendant health issues. The Canton's Civil Protection Administration received Gap funding to conduct the building-level energy audits preparatory to developing ameliorative projects.

11.5.4 *Establishing Municipal Creditworthiness*

Between 2013 and 2023, the World Bank sponsored the City Creditworthiness Initiative for the municipal officials from the Global South. Supported by the Rockefeller Foundation, the Korean Green Growth Trust Fund, and others, the program included instituting Creditworthy Academies (offering three- to five-day courses), multi-year follow-up technical assistance and mentoring of graduates, and knowledge products, including a local government borrowing database and a creditworthiness self-assessment tool kit.

11.5.5 *Brazil's PAC Program*

Originally conceived during President Lula's first term in 2007, the Program to Accelerate Growth (PAC), it is now reorganized as NOVO PAC, with a $1.7 trillion budget to invest in a wide range of physical and social infrastructure throughout the country. It explicitly depends on multi-level governance involving national, state, and city governments in the projects. It has a special carve-out of about $100 billion for sustainable and resilient cities. Some of the funding is for urban adaptation and flood control, divided between drainage and slope containment. This program is financing projects in 235 municipalities in 26 states, including drainage projects in the recently flooded Porto Alegre and slope containment in Rio de Janeiro's favelas (Agencia Gov, 2025).

11.5.6 *Melbourne: 20-Minute City*

In its plan for 2017–2050, revised in 2019, Melbourne outlined its support of 20-minute neighborhoods. Based on a whole-of-government approach, this plan not only embeds the concept in infrastructure plans but also makes it the centerpiece of how it will accommodate future demographic trends sustainably. It specifies the contents of a

20-minute neighborhood based on the accessibility within a 20-minute walk of amenities such as shopping, health facilities, recreation, affordable housing options, and schools, as well as the ability to age in place and access to local public transport including safe cycling networks, and life-long learning opportunities. These compact-development ideas yield climate policies with multiple co-benefits.

11.6 A Research Agenda for System Transformation

This review of the current state of urban climate finance indicates that a good deal of new knowledge has been produced in the past few years. These insights, drawn from practice and from the emergence or consolidation of new sources of data, have in turn, informed practice in a virtuous circle. Practice has also incorporated innovations now informing research. These efforts are beginning to answer some of the questions outlined in the updated version of the GRAA. Emerging theories are ready for testing. Yet as explored within the framework of this chapter, many areas for future research have emerged. The overriding need is contextualization: detailed country-level studies that answer a multitude of logistical and substantive questions about how to conceptualize, manage, execute, and evaluate systems approaches to climate finance, governance, and partnerships. The list that follows, which is not comprehensive, but suggestive, identifies the need for a better understanding of:

- How to prioritize investments in cities of differing sizes, geographies, capacities, and risk profiles within a development framework and urban plan; how to balance citizen participation with expert knowledge about such prioritizations; and how to develop scenarios to assist in decision-making about policy and program selection at local and higher levels of government, including how to collect the data, especially those measuring the climate effectiveness, cost benefits, and return on investment of various constellations of urban-systems approaches.
- How to put into practice and manage a systems approach in cities, including understanding which systems can be developed together or in parallel; and how to measure, and then achieve, maximum

climate goals for mitigation and/or adaptation within a city or region through a tailored systems approach.

- How to align national climate goals with subnational ones. For example, should countries decide to develop country platforms, how they would include urban interests: would they prioritize the interests of a particular group of cities due to the impact on climate goals, or would they use some other criteria? What responsibilities would each sphere of government have in the implementation of the country platform elements and how would national governments manage urban financial needs—would they increase flows directly to cities or have their relevant ministries undertake the tasks? In these cases, how would the different responsibilities be coordinated and monitored?
- How and by what means countries can increase their cities' creditworthiness and capacities; how national and subnational governments can agree on the amount and type of debt a city might take on.
- Detailed legal provisions by country of the respective enabling environments in which national and subnational governments would link their financing and funding schemes for climate investments. Of interest are a group of questions related to subnational competencies and the management of subnations debt. For example, under what conditions can a city legally borrow and service its debt? Protect its creditors in the event of political change? Execute a project in a timely fashion? What recourse exists if things go wrong? Answers to these questions are followed by another set: what amendments to national law in each country would be needed to establish means for cities to de-risk public and private urban-systems investments?
- More details about scaling up through aggregating investments among cities are needed. For example, what types of infrastructure lend themselves to packaging? What would be the desirable overall ticket size of this kind of project? How would the project be managed; that is, how would funds be distributed and monitored? What would happen if one or more of the participant cities was non-performing?
- The nature, type, composition, and size of partnerships needed to carry out the various recommendations for a systems approach to urban climate finance, including how to create and manage financed projects and how to develop trust and respect among the partners.

- How artificial intelligence can be of service in responding to these questions.

11.7 Conclusion

This discussion of urban-systems approaches to finance, governance, and multi-level partnerships covers several themes. First, it establishes a monetary target for needed investments in key infrastructure systems based on existing financial flows and estimates of gaps. It argues that for maximum effect, urban climate investment should focus on urban systems, not individual sectors. Second, it explains an urban-systems approach and identifies the key infrastructure investments. It outlines the requirements to scale up investments, including engaging in urban planning, moving away from project-by- project financing to programmatic, designing and implementing transformative interventions, and aligning supportive, enabling environments with the capacity to accommodate both financing and funding, including creditworthiness and other factors. Third, while the discussion establishes how and under what conditions cities are critical to achieving global climate goals, it also argues that cities cannot do it alone. Cities must work in partnership with all spheres of government, as acknowledged in the Paris Agreement and subsequent COP declarations. It holds that translating these acknowledgements into investments calls for effective multi-level governance that distributes the roles, responsibilities, and power to the government units most appropriate to achieve the necessary tasks and which have the political will and skills to do so. Fourth, it provides examples from around the world of finance instruments, technical and financial assistance, and programs that encompass an urban-systems approach and multi-level governance. It concludes that today, these ideas are mainly aspirational, as future iterations of the GRAA will identify remaining or emergent research gaps and priority actions needed in response, detailing how such approaches would work at all levels of government.

Competing Interests The authors have no conflicts of interest to declare that are relevant to the content of this chapter.

References

Agencia. (2025). *Novo PAC*. https://www.gov.br/casacivil/pt-br/novopac/cidades-sustentaveis-e-resilientes. Accessed February 19, 2026.

Angel, S., Blei, A., Parent, L., Lamson-Hall, P., & Sánchez, N. (2016). *Atlas of urban expansion—2016 edition*. https://www.lincolninst.edu/publications/other/atlas-urban-expansion-2016-edition/. Accessed February 19, 2026.

AON. (2023). *Weather, climate and catastrophe insight*. https://www.aon.com/getmedia/f34ec133-3175-406c-9e0b-25cea768c5cf/20230125-weather-climate-catastrophe-insight.pdf. Accessed February 19, 2026.

Behnisch, M., Kruger, T., & Jaeger, J. (2022). Rapid rise in urban sprawl: Global hotspots and trends since 1990. *Plos Sustainability and Transformation, 1*(11). https://doi.org/10.1371/journal.pstr.0000034

Boex, J., Boukerche, S., Mason, D., & Whittington, J. (2021). *The state of cities climate finance part 2: The enabling conditions for mobilizing urban climate finance (English)*. World Bank Group. http://documents.worldbank.org/curated/en/602521626243370465. Accessed February 19, 2026.

CCFLA & SDSN Commission. (2025). *Integrating urban and subnational priorities into country platforms: Policy brief*. https://citiesclimatefinance.org/publications/integrating-urban-and-subnational-priorities-into-country-platforms. Accessed February 19, 2026.

CCFLA. (2024). *The state of cities climate finance 2024*. https://citiesclimatefinance.org/publications/2024-state-of-cities-climate-finance. Accessed February 19, 2026.

CHAMP. (2025). *What is CHAMP?* https://www.champ-climate.org/. Accessed February 19, 2026.

Climate Bond Initiative. (2023). *Financial aggregation blueprints for urban climate infrastructure—Annex 2: Municipal green bonds blueprint*. https://www.climatepolicyinitiative.org/wp-content/uploads/2023/06/Annex-2-Municipal-Green-Bonds-Blueprints.pdf. Accessed February 19, 2026.

Climate Bond Initiative. (2025). *Global State of the Market 2024*. https://www.climatebonds.net/files/documents/publications/Climate-Bonds_Sustainable_Debt_2024_Global-State-of-the-Market_24-Sep-2025.pdf. Accessed February 19, 2026.

Coalition for Urban Transitions. (2019). *Climate emergency, urban opportunity: How national governments can secure economic prosperity and climate catastrophe by transforming cities*. https://urbantransitions.global/wp-content/uploads/2019/09/Climate-Emergency-Urban-Opportunity-report.pdf. Accessed February 19, 2026.

COP 27 President. (2023a). Matters relating to adaptation, Glasgow-Sharm el-Sheikh work program on the global goal on adaptation referred to in decision 7/CMA.3 (Draft). https://unfccc.int/sites/default/files/resource/Global_goal_on_adaptation_1.pdf. Accessed February 19, 2026.

COP 27 President. (2023b). *SURGe sustainable urban resilience for the next generation*. https://unhabitat.org/sites/default/files/2023/06/cop27_surge_initiative_updated_concept_note.pdf. Accessed February 19, 2026.

COP 28 President. (2023). *Local climate action summit*. https://www.cop28.com/en/lcas. Accessed February 19, 2026.

COP 29 President. (2024). *Declaration on multisectoral actions pathways (MAP) to resilient and healthy cities*. https://cop29.az/en/pages/cop29-declaration-on-multisectoral-actions-pathways-map-to-resilient-and-healthy-cities. Accessed February 19, 2026.

Currimbhoy, S. (2025, September). *Under pressure, overdue: The portfolio approach and financing cities for resilience*. Resilient Cities Network RCN. https://resilientcitiesnetwork.org/wp-content/uploads/2025/10/Resilience-Finance-Under-Pressure-Overdue.pdf. Accessed February 19, 2026.

Deuskar, C., Murray, S., Molano, J., Khan, A., & Augustin, M. (2025). *Banking on cities: Investing in resilient and low-carbon urbanization*. Urban Development Series. World Bank. https://openknowledge.worldbank.org/server/api/core/bitstreams/71fce774-9761-4718-864b-50baf35ee612/content. Accessed February 19, 2026.

Dodman, D., Hayward, B., Pelling, M., Castan Broto, V., Chow, V., Chu, E., Dawson, R., Khirfan, L., McPhearson, T., Prakash, A., Zheng, Y., & Ziervogel, G. (2022). Cities, settlements and key infrastructure. In H.-O. Pörtner, D. C. Roberts, M. Tignor, E. S. Poloczanska, K. Mintenbeck, A. Alegría, M. Craig, S. Langsdorf, S. Löschke, V. Möller, A. Okem, & B. Rama (Eds.), *Climate change 2022: Impacts, adaptation and vulnerability. Contribution of working group II to the sixth assessment report of the intergovernmental panel on climate change*. Cambridge University Press. https://doi.org/10.1017/9781009325844.008

Frost, R. (2024, September 5). Severe flooding leaves at least 100 dead and thousands homeless in Brazil. *Euronews*. https://www.euronews.com/green/2024/05/09/severe-flooding-leaves-at-least-100-dead-and-thousands-homeless-in-brazil#:~:text=Green%20Green%20News-,Severe%20flooding%20leaves%20at%20least%20100%20dead%20and%20thousands%20homeless,200%2C000%20people%20have%20been%20displaced. Accessed February 19, 2026.

Gilmour, A., Tanaka, J., & Colenbrander, S. (2024). *Designing and governing country platforms: What role for the MDB's*. ODI Report. https://media.odi.org/documents/Designing_and_governing_country_platforms.pdf. Accessed February 19, 2026.

GIZ. (2018). *Multi-level climate governance supporting local action*. https://www.citiesalliance.org/sites/default/files/giz2018-0318en-cpmud-multi-level-climate-governance.pdf. Accessed February 19, 2026.

Global Covenant of Mayors. (2024). *Banking on change, finance for local climate action.* https://www.globalcovenantofmayors.org/wp-content/uploads/2024/11/GCoM-Impact-Report-24.pdf. Accessed February 19, 2026.

Heubaum, H., Brandon, C., Tanner, T., Surminski, S., & Roezer, V. (2022). *The triple dividend of building climate resilience: Taking stock, moving forward.* World Resources Institute Working Paper. https://doi.org/10.46830/wriwp.21.00154

Instituto Nacional de Ecología y Cambio Climático. (2018, August 20). *México comprometido con sus CND hacia una economía verde y baja en carbono.* Gobierno de Mexico. https://www.gob.mx/inecc/prensa/mexico-comprometido-con-sus-cnd-hacia-una-economia-verde-y-baja-en-carbono. Accessed February 19, 2026.

Irvin, A., Oke, C., Khan, S., Jance IV, B., Ranalder, L., Barth, B., Deacon, A., Abusamra, R., Addison, A., Adebisi, J., Adsett, C., Ailio, E., Al Jadaa, D., Aslan, S., Assalini, S., Assarkhaniki, Z., Astolfo, J. P., Badino, M., Bai, X., …, & Williams, A. (2024). *Next-generation city climate action: Findings from the 2024 Innovate4Cities conference and update to the global research and action agenda for cities and climate change science (GRAA).* Global Covenant of Mayors for Climate and Energy, UN-Habitat & University of Melbourne. https://doi.org/10.26188/27147612

Lwasa, S., Seto, K. C., Bai, X., Blanco, H., Gurney, K. R., Kılkış, Ş., Lucon, O., Murakami, J., Pan, J., Sharifi, A., & Yamagata, Y. (2022). Urban systems and other settlements. In P. R. Shukla, J. Skea, R. Slade, A. Al Khourdajie, R. van Diemen, D. McCollum, M. Pathak, S. Some, P. Vyas, R. Fradera, M. Belkacemi, A. Hasija, G. Lisboa, S. Luz, J. & Malley (Eds.), *IPCC, 2022: Climate change 2022: Mitigation of climate change. Contribution of working group III to the sixth assessment report of the intergovernmental Panel on climate change.* Cambridge University Press. https://doi.org/10.1017/9781009157926.010

Maciel, E. (2024, May 22). Rebuilding a state devastated by flood: Brazil's Investment and Solidarity Initiative. *Development Aid.* https://www.developmentaid.org/news-stream/post/179591/rebuilding-brazil-devastated-by-floods. Accessed February 20, 2026.

Mahendra, A., & Setom K. (2019). *Upward and outward growth: Managing urban expansion for more equitable cities in the Global South.* Working Paper. World Resources Institute. https://www.wri.org/research/upward-and-outward-growth-managing-urban-expansion-more-equitable-cities-global-south. Accessed February 20, 2026.

Mahendra, A., King, R., Du, J., Dasgupta, A., Beard, V. A., Kallergis, A., & Schalch, K. (2021). *Seven transformations for more equitable and sustainable cities.* World Resources Report, *Towards a More Equal City.* World Resources Institute. https://doi.org/10.46830/wrirpt.19.00124

MDB Climate Action Group. (2024). *Country platforms for climate action: MDB statement on understanding and way forward*. https://documents1.worldbank.org/curated/en/099427511122436202/pdf/IDU-eb55cea2-53b2-4e2d-88cd-6a23e8b08a0a.pdf. Accessed February 20, 2026.

Melbourne. (n.d.) *20-minute neighbourhood projects*. https://www.planning.vic.gov.au/guides-and-resources/strategies-and-initiatives/20-minute-neighbourhoods/20-minute-neighbourhood-projects. Accessed February 20, 2026.

Michaelides, K., Singer, M., & Fox, S. (2025). *Water and climate: Rising risks for urban populations*. University of Bristol, University of Cardiff. https://washmatters.wateraid.org/sites/g/files/jkxoof256/files/2025-03/Water-and-climate-Rising-risks-for-urban-populations.pdf. Accessed February 20, 2026.

Michelon, W., Santos, L., Santin, L., & Viancelli, A. (2025). Increased leptospirosis incidence following flooding in Rio Grande do Sul, Brazil. *Journal of Environmental Health Research, 1–6*. https://doi.org/10.1080/09603123.2025.2498623

Molina, L. (2025, May 6). *Floods in Rio Grande do Sul exposed the climate crisis*. https://cop30.br/en/news-about-cop30/floods-in-rio-grande-do-sul-exposed-the-climate-crisis. Accessed February 20, 2026.

Mylnar, M., & Belmonte, J. (2024, February 29). Cities are the missing piece for countries to address climate change. *The Economist*. https://impact.economist.com/sustainability/resilience-and-adaptation/cities-are-the-missing-piece-for-countries-to-address-climate-change. Accessed February 20, 2026.

Pinko, N., Negreiros, P., Wetherbee, C., & Tonkonogy, B. (2022, May). *Financial aggregation for cities*. https://www.climatepolicyinitiative.org/wp-content/uploads/2022/06/Financial-Aggregation-for-Cities.pdf. Accessed February 20, 2026.

Ranalder, L., Tollin, N., Gragnani, P. & Geoghegan, P. (2025). *Seizing the urban opportunity: A 3-step guide for strengthening Nationally Determined Contributions (NDC 3.0)*. United Nations Human Settlements Programme (UN-Habitat). https://unhabitat.org/seizing-the-urban-opportunity-a-3-step-guide-for-strengthening-nationally-determined-contributions. Accessed February 20, 2026.

Regan, H. (2024, June 23). Nearly 2 million people stranded as second wave of devastating floods hits Bangladesh in less than a month. *CNN*. https://www.cnn.com/2024/06/22/asia/bangladesh-flooding-sylhet-climate-intl-hnk. Accessed February 20, 2026.

Revi, A., Roberts, D., Klaus, I., Bazaz, A., Krishnaswamy, J., Singh, C., Eichel, A., Kodira, P., Seth, S., Adelekan, I., Babiker, M., Bertoldi, P., Cartwright, A., Chow, W., Colenbrander, S., Creutzig, F., Dawson, R., De Coninck, H.,

De Kleijne, K., ..., & Ürge-Vorsatz, D. (2022). *The summary for urban policymakers of the IPCC's sixth assessment report.* Indian Institute for Human Settlements. https://doi.org/10.24943/SUPSV511.2022

Rodrigues, M. (2024). Lack of maintenance and design flaws prevented the flood containment system from Protecting Porto Alegre. *Pesquisa FAPESP.* 341 (July). https://revistapesquisa.fapesp.br/en/lack-of-maintenance-and-design-flaws-prevented-the-flood-containment-system-from-protecting-porto-alegre/. Accessed February 20, 2026.

Schipper, E., Revi, A., Preston, B., Carr, E., Eriksen, H., Fernandez-Carril, L. Glavovic, C., Hilmi, H., Ley, R., Mukerji, M., Muylaert de Araujo, R., Perez, S., & Singh, P. (2022). Climate resilient development pathways. In *Climate change 2022: Impacts, adaptation and vulnerability. contribution of working group II to the sixth assessment report of the intergovernmental panel on climate change* (pp. 2655–2807). Cambridge University Press. https://doi.org/10.1017/9781009325844.027

Simoes-Sousa, I., Camargo, M., Tavora, J., Piffer-Braga, A., Farrar, J., & Pavelsky, T. (2025). The May 2024 flood disaster in southern Brazil: Causes, impacts, and SWOT-based volume estimation. *Geophysical Research Letters, 52*(4). https://doi.org/10.1029/2024GL112442

Tanner, T., Surminski, S., Wilkinson, E., Reid, R., Rentschler, J., & Rajput, S. (2019). *The triple dividend of resilience: Realising development goals through the multiple benefits of disaster risk management. Global Facility for Disaster Reduction and Recovery (GFDRR) at the World Bank and Overseas Development Institute (ODI).* https://www.odi.org/publications/9599-triple-dividend-resilience-development-goals-multiple-benefits-disaster-risk-management. Accessed February 20, 2026.

UNFCCC. (2024). Report of the conference of the parties serving as the meeting of the parties to the Paris agreement on its fifth session, held in the United Arab Emirates from 30 November to 13 December 2023. UN doc FCCC/PA/CMA/2013/16. https://unfccc.int/sites/default/files/resource/cma2023_16E.pdf. Accessed February 20, 2026.

United States Environmental Protection Agency. (2025, March). *About small wastewater systems.* https://www.epa.gov/small-and-rural-wastewater-systems/about-small-wastewater-systems#:~:text=Many%20small%20and%20rural%20communities,Small%20Communities%20Challenges%20and%20Needs. Accessed February 20, 2026.

UN-Habitat, SDU Resilience, & UNDP. (2024). Urban content of NDCs: Local climate action explored through in-depth country analysis 2024 Report. Global Covenant of Mayors for Climate & Energy. [Lead authors: Nicola Tollin, James Vener, Yu Liu, Patrizia Gragnani, Maria Pizzorni, and Bernhard Barth]. https://unhabitat.org/sites/default/files/2025/09/ndc_global_report_2024.pdf. Accessed February 19, 2026.

White, R., & Wahba, S. (2019). Addressing constraints to private financing of urban (climate) infrastructure in developing countries. *International Journal of Urban Sustainable Development*, *11*(3), 245–256. https://doi.org/10.1080/19463138.2018.1559970

World Bank. (2015). *City creditworthiness initiative: A partnership to deliver municipal finance*. https://www.worldbank.org/en/topic/urbandevelopment/brief/city-creditworthiness-initiative. Accessed February 20, 2026.

World Economic Forum. (2022, August).*Delivering climate-resilient cities using a systems approach*. Global Future Council on Cities of Tomorrow. https://www3.weforum.org/docs/WEF_C4IR_GFC_on_Cities_Climate_Resilience_2022.pdf. Accessed February 20, 2026.

CHAPTER 12

Conclusion: Toward Climate-Resilient Development in Cities

Cathy Oke, Nicola Tollin, Lea Ranalder, Benjamin Jance IV, Debra C. Roberts, Marnie McGregor, Aphrodite Bouikidis, Meggan Spires, and Juliet Oluoch

12.1 Introduction

Cities stand at a critical juncture in human history. Housing over half the world's population while generating 67–72% of global greenhouse gas emissions, urban areas simultaneously represent humanity's greatest climate challenge and most promising opportunity for transformative action (Lwasa et al., 2022; Vogel et al., 2024). The preceding chapters have documented a journey—from the 2016 proposal for the IPCC

C. Oke (✉)
Melbourne Centre for Cities, Faculty of Architecture Building and Planning, University of Melbourne, Melbourne, Australia
e-mail: cathy.oke@unimelb.edu.au

N. Tollin
UNESCO Chair on Urban Resilience, University of Southern Denmark, Odense, Denmark
e-mail: nto@iti.sdu.dk

C. Oke et al. (eds.), *A Global Research and Action Agenda on Cities and Climate Change Science*,
https://doi.org/10.1007/978-981-95-6517-7_12

Special Report on Climate Change and Cities (SRCCC) (Chapter 2) through three iterations of the Global Research and Action Agenda on Cities and Climate Change Science (GRAA) (Chapter 3)—demonstrating how sustained collaborative effort can bridge traditionally isolated communities of science, policy, and practice (Hunter et al., 2022).

This chapter distills insights from the Innovate4Cities initiative's evolution (Chapters 2–3) and the systems approaches to urban research and innovation (Chapters 4–11). While cities have mobilized unprecedented climate ambition, with over 13,800 municipalities making a commitment to action through the Global Covenant of Mayors for Climate & Energy (GCoM, 2024), critical implementation gaps persist. Current

L. Ranalder
UN-Habitat, Nairobi, Kenya
e-mail: lea.ranalder@un.org

B. Jance IV · M. McGregor
Global Covenant of Mayors for Climate and Energy, Brussels, Belgium
e-mail: bjance@globalcovenantofmayors.org

M. McGregor
e-mail: mmcgregor@globalcovenantofmayors.org

D. C. Roberts
School of Life Sciences, University of KwaZulu-Natal, Durban, South Africa

Faculty of Geo-Information Science and Earth Observation, University of Twente, Enschede, The Netherlands

D. C. Roberts
e-mail: Debra.Roberts@palebludot.co.za

A. Bouikidis
Independent (Urban Resilience Specialist), Athens, Greece
e-mail: aphrodite.bouikidis@gmail.com

M. Spires
ICLEI Africa, Cape Town, South Africa
e-mail: meggan.spires@iclei.org

J. Oluoch
Ufanisi Research Network, Nairobi, Kenya
e-mail: oluochjuliet08@gmail.com

urban climate finance is far below estimated needs: IPCC WGIII reports about US$384 billion per year (2017–2018), roughly 10% of the investment required for low-carbon, resilient urban development, while more recent tracking places urban flows at about US$831 billion (2021–2022) versus estimated city needs of ~US$4.5 trillion per year, a persistent financing gap concentrated in rapidly urbanizing regions of the Global South (IPCC, 2022; CCFLA/CPI, 2024; Chapter 8). Moreover, the crises confronting cities extend beyond climate change alone. Climate risk can have multiplying effects on other urban challenges, such as air quality, housing density, and health-care access, demanding integrated approaches (Mahadevia et al., 2025; Chapter 2). Meanwhile, inequality continues widening, greenhouse gas emissions are reaching record levels (IEA, 2025), and the increasing likelihood of exceeding the 1.5 °C Paris Agreement target necessitates urgent acceleration of both climate mitigation and adaptation (Bevacqua et al., 2025; Cannon, 2025; Gasser et al., 2015; Kirchengast & Pichler, 2025; Reisinger et al., 2025; Theokritoff et al., 2025).

This chapter brings together key findings across the three dimensions of the book's structure: (1) the evolution of the knowledge co-production frameworks enabling science–policy–practice collaboration (Chapters 2–3); (2) operational mechanisms facilitating two-way translation between research and local action through boundary spanning, networks, and justice-centered approaches (Chapters 4–8); and (3) systems perspectives integrating finance, governance, infrastructure, biodiversity, health, and culture (Chapters 9–11). We frame these insights against the imperative for urban transformation through climate-resilient development, which is defined by the IPCC as processes implementing ambitious mitigation and adaptation to support sustainable development while meeting basic human needs, eliminating poverty, enhancing biodiversity, and advancing justice and equity (Revi et al., 2022).

This chapter concludes with action-oriented recommendations for urban climate researchers, practitioners, and policymakers, alongside identification of critical research priorities that could inform the IPCC SRCCC, due 2027, imperatives for future Innovate4Cities activity, and any further updates of the GRAA (Sect. 12.3).

12.2 Synthesizing Key Findings

12.2.1 The Evolution of Collaborative Knowledge Production (Chapters 2–3)

The Innovate4Cities (2018–2021–2024) initiative has pioneered a model for bridging city-centered disciplinary and sectoral boundaries. From an initial 700 participants to nearly 2000, alongside the evolution of the GRAA from 80 to 160 knowledge gaps and 259 action priorities, Innovate4Cities has introduced methodological innovations and improved the understanding of how cities can localize responses to global climate challenges (Irvin et al., 2024; Chapter 3; Prieur-Richard et al., 2019).

Co-production Processes have integrated diverse knowledge systems: The 2018 Cities and Climate Change Science Conference (the 2018 Conference), laid the early groundwork for multi-stakeholder collaboration by convening researchers, practitioners, and city leaders to define priority knowledge gaps for urban climate science (Prieur-Richard et al., 2019; Chapter 2). These processes directly informed the later Innovate4Cities conferences (2021, 2024), which expanded this foundation by operationalizing co-production, integrating diverse knowledge systems, and updating the Global Research and Action Agenda. By 2024, digital-collaboration tools, including real-time Miro boards enabling dynamic, iterative agenda development that transcends traditional paper-based collection methods (Irvin et al., 2024; Chapter 3).

Boundary-Spanning Mechanisms have facilitated knowledge translation, foresight, and "horizon scanning" for relevant challenges and priorities across organizational and disciplinary divides (Chapter 4): GCoM's Research and Innovation Technical Working Group and the Innovate4Cities Conference Advisory Committee have engaged diverse stakeholders across local and national government, academia, industry, and civil society to support ongoing science–practice dialogue (Oke et al., 2022; Chapter 2). Innovate4Cities Marketplaces highlight the opportunity for deeper cross-sector collaboration, especially when matching city practitioner and policy demands with solutions and knowledge from businesses, researchers, and others (Jance et al., 2024).

Justice and Equity have evolved from peripheral consideration to foundational principle: The 2021 GRAA elevated justice and equity to a cross-cutting theme (Walsh et al., 2022; Chapter 2), and by 2024, the importance of justice and equity for effective climate action had been further recognized by the inclusion in the GRAA "building" foundation. (Irvin et al., 2024; Chapter 3). Case studies around the world show how place-based approaches centering marginalized communities generate more effective and durable outcomes (Chapter 7).

12.2.2 *Delivery Approaches: From Knowledge to Action (Chapters 5–8)*

City and science networks function as critical knowledge brokers to catalyze *long-term partnerships and co-design processes* (Chapters 5, 6): C40 Cities, ICLEI, UCLG, the Urban Climate Change Research Network (UCCRN), and the GCoM alliance provide platforms for peer-to-peer exchange, pilot programs to test innovations, and access to implementation insights (Acuto & Leffel, 2020; Bulkeley & Betsill, 2013; Fünfgeld, 2015). These networks demonstrate the capacity to rapidly disseminate evidence-based practices (City of Freetown, 2023; Municipality of Lima, 2024; Chapter 5), and the incorporation of Indigenous-municipal collaborations (Bush et al., 2023; Chapter 6) exemplify data-driven interventions enabled through network support.

Data and digitalization strategies enhance transparency, accountability, and responsiveness (Chapter 8): Granular, place-specific climate projections and vulnerability assessments enable targeted interventions (Creutzig et al., 2019), while the emergence of artificial intelligence (AI) presents opportunities to generate efficiencies in data-intensive city climate-action planning processes (Irvin et al., 2024). However, digital divides and surveillance concerns necessitate governance frameworks that embed ethics-by-design principles (Kitchin et al., 2015; Chapter 8). Balancing technological optimization and fair and just participation requires further guidance.

Analysis across case studies reveals four critical dimensions for embedding *justice and equity in climate action*. (1) *Procedural justice* ensures meaningful participation in decision-making (Chapters 7, 9). However, successful co-production demands that inherent power imbalances between communities and state institutions be addressed (Joshi & Moore, 2004; Shand, 2018). (2) *Distributive justice* addresses fair allocation of climate action benefits and burdens (Chapters 7, 9). Yet green gentrification risks demand-proactive safeguards (Anguelovski & Connolly, 2022; Cucca et al., 2023). (3) *Recognition justice* validates diverse knowledge systems and cultural practices. The integration of

Indigenous Knowledge and Local Knowledge systems into planning processes, while progressing, remains incomplete, and future iterations require strengthening mechanisms for systematic incorporation of diverse epistemologies beyond tokenistic consultation (Gazing Wolf et al., 2024; Orlove et al., 2023; Chapters 6, 10). (4) *A capabilities approach* focuses on expanding freedoms and opportunities, demonstrating how climate action can simultaneously enhance human capabilities and community autonomy (Quezon City Government, 2022; Chapter 7).

The scale of *urban informality* underscores the centrality of justice and equity in climate action. The 2018 Cities and Climate Change Science Conference gave prominent attention to informality, establishing a critical link between climate change and the lived reality of billions of urban residents (Prieur-Richard et al., 2019; UN-Habitat, 2018). With UN-Habitat projecting that informal settlements could house 3 billion people by 2050 (UN-Habitat, 2024), this represents massive vulnerability and exposure as cities confront climate overshoot. The GRAA's evolution demonstrates that addressing informality requires more than infrastructure: the formal sector must embrace sufficiency principles (Chapter 10), decolonize urban development approaches (Chapters 6, 9), prioritize community-led systems (Chapter 7), and embed informal settlement dwellers' perspectives within formal decision-making (Almansi et al., 2011; Shand, 2018). However, informality has declined in GRAA prominence since 2018, and the current building analogy may inadvertently marginalize informal systems. Continuing to evolve the GRAA's representation to highlight applicability across all urban challenges, including informality, remains essential for ensuring climate action reaches the world's most vulnerable populations.

12.2.3 Systems Approaches: Biodiversity, Culture, Geography, Infrastructure, Finance, and Governance (Chapters 9–11)

Urban form and polycentric development have emerged as critical mitigation and adaptation levers (Chapter 10). The shift toward proximity-based models demonstrates how compact, mixed-use development simultaneously reduces transport emissions, enhances service access, and improves livability, transcending density alone and prioritizing pedestrian-friendly environments and ecological connectivity (Van Nes, 2021; Chapter 10). However, implementation demands equity safeguards that address displacement risks and ensure affordable housing within transit-oriented development. Applying these principles and adapting them to Global South and rapidly urbanizing contexts represents both knowledge gaps and action priorities that need to be addressed. *Circular economies and resource efficiency* prove essential, given cities' consumption of 75% of global material resources (Chapter 10). Advancing circularity at scale requires coherent multi-level frameworks, updated building codes, fiscal tools, and systematic urban metabolic modeling to inform accountability (UNEP-UN-Habitat, 2021; Chapter 10). *Social infrastructure* extends beyond physical systems to encompass libraries, parks, care centers, and community hubs as "essential foundations of resilience, equity, justice and democratic social cohesion" (Hollis et al., 2023; Chapter 10). *Digital technologies and AI* offer transformative potential that needs to remain inclusive and responsible (Chapters 8, 10). However, digital divides limiting access, risks of bias in models trained predominantly on Global North data, and significant energy and water footprints associated with AI infrastructure (Shi et al., 2023; Simon & O'Brien, 2023) necessitate governance and operational frameworks ensuring equitable access and sustainable deployment. *Finance and multi-level governance* remain critical enablers and barriers (Chapter 11). Annual urban climate finance reached US$831 billion in 2021–2022, yet falls dramatically short of the US$4.5 trillion needed by 2030 (CCFLA, 2024), with low- and middle-income countries facing acute gaps, requiring US$256–821 billion annually through 2050 (CCFLA, 2024; Deuskar et al., 2025; Chapter 11). Current project pipelines show 86% of reported urban climate projects seeking financing, with 48% completely unfunded; emerging market and developing economies face particularly acute challenges, with 40% of projects unfunded compared to 25% in developed economies (Chapter 11). The Coalition for High Ambition Multi-level Partnerships (CHAMP) demonstrates promising governance reforms (ICLEI, 2024; Chapter 11). Analysis reveals that adequately supported city action could close approximately 40% of the emissions gap between current nationally determined contributions (NDCs) and Paris-aligned trajectories (C40 & GCoM, 2025; Chapter 11).

Biodiversity-culture-geography nexus demands integrated frameworks (Chapter 9): Geography fundamentally shapes each city's climate challenges and opportunities (Castree et al., 2025). Urban biodiversity faces compounding threats from climate change, invasive species, habitat fragmentation, and pollution (IPBES, 2019; Pörtner et al., 2021;

Chapter 10), while simultaneously providing critical ecosystem services including heat mitigation, stormwater management, and air-quality improvement (Elmqvist et al., 2013; Chapter 10). The integration of cultural considerations—including Indigenous Knowledge systems, multispecies justice frameworks, and community-engagement processes—proves essential for contextually appropriate, durable responses (Celermajer et al., 2021; Waterton, 2015; Chapter 10).

12.3 Action-Oriented Recommendations From Innovate4cities

This book illustrates how the Innovate4Cities process serves as a series of collaborative, multi-level, and cross-sector opportunities to measure progress against known city-related climate-knowledge gaps and action priorities, as well as a "radar" that generates critical foresight into topics, levers, mechanisms, and knowledge opportunities that wield potential for progress. Building upon synthesized findings, we present targeted recommendations across stakeholder groups, recognizing that transformative change requires coordinated action that spans scales and sectors.

12.3.1 For Researchers and Academic Institutions

Prioritize Transdisciplinary Collaboration: Universities should establish dedicated urban-climate research centers, building upon collaborative models pioneered by Innovate4Cities (McPhearson et al., 2016; Chapter 4). Strengthening science–policy–practice partnerships between academia, governments, civil society, and industry ensures that co-produced knowledge is action-oriented and has impact. Such collaboration enables co-creation, brokerage, and mutual learning that enhances societal uptake and transformative potential (Chapter 4).

Localize Research Agendas: The persistent Global South research gap demands proactive correction (Kendal et al., 2020; Chapter 9). Research should prioritize place-based challenges while building local capacity through citizen-science initiatives combining data collection with diverse knowledge systems (Tengö et al., 2021; Chapter 9). Research must be conducted with and for society, embedding community participation throughout the research cycle to inform decision-making and drive equitable implementation.

Advance Solution-Oriented Enquiry: Progress requires shifting toward solution-oriented inquiry that examines how to implement and accelerate climate solutions in diverse contexts (Solecki et al., 2021; Chapter 2). This includes building durable local capacities, responsibly integrating appropriate technologies, and enabling institutional transformations. Strengthening knowledge transfer and brokerage mechanisms through practice-based research and translational platforms can help bridge the persistent divide between scientific insight and policy–practice action.

Strengthen Indigenous Knowledge and Local Knowledge Integration: Systematic incorporation of Indigenous Knowledge systems into urban climate research remains incomplete (Bartlett et al., 2012; Nelson & Shilling, 2018; Chapters 6, 10). Future work should move beyond tokenistic consultation toward genuine co-production that recognizes diverse epistemologies as equally valid knowledge sources. Community knowledge is a vital resource, and research must offer genuine value in return, ensuring that community efforts lead to tangible outcomes and sustained engagement.

Develop methodologies for linking climate shock impacts to investment requirements at city and systems levels, collecting disaggregated data to measure infrastructure investment effectiveness and ROI (Chapter 11).

12.3.2 For Urban Practitioners and City Networks and Initiatives

Address Finance Gaps Through Diversified Mechanisms: Strengthening municipal creditworthiness is essential for accessing capital markets and innovative financing instruments (Chapter 11). Cities should maximize revenues through efficient tax collection and utility billing while exploring structured and blended finance models, especially those that focus on de-risking municipal investment. Financial resources for implementation can only be unlocked when cities have the necessary data, and institutional capacities are critical means of implementation. Robust data systems enable cities to assess climate risks, identify bankable projects, and track performance, enhancing investor confidence and accountability. Sustained capacity development through technical training, peer-to-peer exchanges, and city network support ensures that local administrations can design, manage, and report on climate investments effectively.

Mainstream Climate Integration: Rather than treating climate change action in isolation, local and national governments must mainstream climate-policy integration across all departments (Bulkeley & Betsill, 2005). This land-use planning, transportation infrastructure, building codes, and water-management strategies to consistently reflect climate change priorities, with the support of aligned budgeting, capacity building, and decision-making routines (Medeiros & van der Zwet, 2020; Chapter 11).

Scale Successful Innovations Strategically: Case studies demonstrate numerous effective local interventions (Table 12.1; Chapters 5, 7, 10, & 11). City climate networks should facilitate systematic documentation, evaluation, and adaptation of proven approaches, recognizing that successful scaling requires contextual translation rather than direct replication. City networks and initiatives need to ensure that climate action is coherently integrated across all levels of advocacy through effective multi-level governance from local communities to regional and national governments (Acuto & Leffel, 2020; Chapter 5).

Table 12.1 Legend:
Timeline* Definitions:

- Short-term (1–3 years): Immediate needs for IPCC SRCCC and near-term policy and practice decisions
- Medium-term (3–5 years): Post-AR7 research agenda building on foundational work
- Long-term (5+ years): Foundational transformative research requiring sustained investment
- Ongoing priority: Continuous attention required throughout all timelines
- *realistically, each of the dimensions will be an ongoing priority.

Stakeholder Categories:

- Researchers: Academic institutions, research centers, IPCC authors
- Practitioners: City governments, urban planners, municipal officials
- Policymakers: National and regional government officials, policy advisors
- Communities: Local groups, Indigenous communities, civil society
- Private-sector entities: industry, business, and investors

Table 12.1 Research-priority matrix—Interconnected themes for consideration for future iterations of GRAA

Priority area	*Key questions*	*Primary methods*	*Timeline*	*Stakeholders*	*Related chapters*
1. *Climate-overshoot implications*					
Climate overshoot implications	What are the impacts of temperature goal overshoot (both exceedance and decline trajectories) and what are the mitigation, adaptation, and development responses?	Systems modeling, vulnerability assessments, scenario analysis	Short-term (1–3 years)	Researchers, practitioners	9, 10, 11
Compound/ cascading risks	How do climate hazards interact with urban vulnerabilities and exposure to generate systemic failures?	Systems modeling, vulnerability assessments, scenario analysis	Short-term (1–3 years)	Researchers, practitioners	8, 10, 11
Adaptation limits	What thresholds trigger the need for transformational vs. incremental change? what adaptation limits will be reached where and when and what this means for determining loss and damage	Comparative case studies, threshold analysis	Medium-term (3–5 years)	Researchers, policymakers	7, 9, 10
Regenerative approaches	How can cities move beyond resilience toward regenerative functions?	Action research, longitudinal studies	Long-term (5+ years)	All stakeholders	9, 10

(continued)

Table 12.1 (continued)

Priority area	*Key questions*	*Primary methods*	*Timeline*	*Stakeholders*	*Related chapters*
2. Localizing knowledge					
Global South research	What context-specific challenges and opportunities exist in rapidly urbanizing regions?	Participatory research, regional assessments	Ongoing priority	Researchers, networks	2, 5, 9
Small and mid-sized cities	How do resource constraints shape climate action in smaller urban centers? How can data and engagement reveal disproportionate risk across small-scale neighborhoods within cities?	Comparative analysis, capacity assessments	Short-term (1–3 years)	Researchers, practitioners	5, 11
Indigenous Knowledge	What methodologies enable genuine co-production with Indigenous communities?	Co-design, two-eyed seeing approaches	Medium-term (3–5 years)	Researchers, communities	6, 9
Cultural indicators	What metrics reflect diverse values beyond standardized frameworks?	Participatory indicator development	Medium-term (3–5 years)	All stakeholders	7, 9
Geographic context and climate vulnerability	How does geographic context (coastal, arid, mountain, polar) fundamentally shape climate vulnerabilities and appropriate solutions?	Comparative geographic analysis, vulnerability assessments, case studies	Ongoing priority	Researchers, practitioners	9 (geography)
Cultural-heritage integration	How can cultural heritage, Indigenous Knowledge, and place identity be meaningfully integrated into climate adaptation and mitigation planning?	Participatory research, case studies, co-design	Medium-term (3–5 years)	Communities, researchers, policymakers	9 (culture), 6

Priority area	*Key questions*	*Primary methods*	*Timeline*	*Stakeholders*	*Related chapters*
3. Implementation Science					
Scaling mechanisms	What conditions enable successful adaptation across contexts?	Process tracing, boundary-spanning, network analysis	Short-term (1–3 years)	Practitioners, networks, industry	4, 5, 9
Capacity-building	What pathways strengthen municipal capabilities most effectively?	Quasi-experimental designs, evaluation studies	Medium-term (3–5 years)	Practitioners, policymakers, industry, networks	11
Behavioral change	What drives shifts in consumption, mobility, and policy support?	Social-psychology research, intervention studies	Medium-term (3–5 years)	All stakeholders	10
Political economy	How do power dynamics shape implementation feasibility?	Political analysis, stakeholder mapping	Short-term (1–3 years)	Policymakers, researchers	7, 11
4. Finance and economics					
Investment needs	What are refined capital requirements by sector/context?	Financial modeling, gap analysis	Short-term (1–3 years)	Policymakers, investors, practitioners, industry	11
Action vs. inaction costs	What are the true economic implications of delayed action?	Cost–benefit analysis, risk assessment	Short-term (1–3 years)	Policymakers, industry, researchers	11
Circular economy	What are the systemic implications of transitioning to circularity?	Systems analysis, case studies	Medium-term (3–5 years)	Practitioners, investors, researchers, industry	10
Creditworthiness	What factors enhance municipal market access?	Financial analysis, institutional assessment	Short-term (1–3 years)	Practitioners, investors, policymakers, industry	11

(continued)

Table 12.1 (continued)

Priority area	*Key questions*	*Primary methods*	*Timeline*	*Stakeholders*	*Related chapters*
Municipal-finance innovation	How can innovative financing instruments (green bonds, blended finance, guarantees) be scaled for urban-climate investment? What are barriers to municipal market access?	Financial analysis, case studies, institutional assessment	Short-term (1–3 years)	Policymakers, investors, practitioners, industry, researchers	11
Climate-finance aggregation	How can cities aggregate small-scale projects to achieve investment scale? What role is there for intermediary institutions?	Financial modeling, institutional analysis, case studies	Short-term (1–3 years)	Practitioners, policymakers, industry, networks	11
Investment ROI by urban system	What are returns on investment for different types of urban climate infrastructure by system (energy, transport, water, buildings)? How can co-benefits be measured?	Cost–benefit analysis, longitudinal studies, systems modeling	Medium-term (3–5 years)	Researchers, investors, policymakers, industry	10, 11
Project pipeline and bankability	What makes urban climate projects bankable? How can pipelines be built from concept to investment-ready?	Financial analysis, case studies, capacity assessments	Short-term (1–3 years)	Practitioners, investors, industry, policymakers	11
5. *Governance*					
Multi-level effectiveness	What mechanisms best coordinate across governance levels?	Institutional analysis, effectiveness evaluations	Medium-term (3–5 years)	Policymakers, networks	5, 11
Policy integration	How to systematically embed climate across sectors?	Policy analysis, mainstreaming studies, boundary-spanning	Medium-term (3–5 years)	Policymakers, practitioners	4, 11
Co-production	What enables meaningful community engagement?	Participatory action research	Ongoing priority	All stakeholders	6, 7, 11

Priority area	*Key questions*	*Primary methods*	*Timeline*	*Stakeholders*	*Related chapters*
Adaptive governance	What frameworks support flexibility under uncertainty?	Adaptive-management research	Medium-term (3–5 years)	Policymakers, researchers	11
Multi-level governance for climate finance	How do multi-level governance structures enable or constrain climate finance flows to cities? What reforms maximize effectiveness?	Policy analysis, institutional assessments, comparative studies	Medium-term (3–5 years)	Policymakers, researchers, networks	11
Private-sector engagement	What partnership models effectively engage private sector in urban climate action? How to align profit motives with climate objectives?	Case studies, institutional analysis, partnership evaluation	Short-term (1–3 years)	Industry, policymakers, practitioners	11
6. Technology and innovation					
AI/digitalization	What are equity implications of digital technologies? What are the resource consumption cost–benefit analysis for various tech deployment?	Technology assessment, equity analysis	Short-term (1–3 years)	Practitioners, researchers, industry, policymakers	8, 10
Nature-based solutions	What is long-term effectiveness under climate change?	Monitoring, experimental studies	Long-term (5+ years)	Practitioners, researchers, communities, industry	9, 10
Infrastructure resilience	How to design for intensifying hazards?	Engineering studies, climate projections	Short-term (1–3 years)	Practitioners, researchers, industry	10, 11
Digital equity and access	How can digital climate tools avoid reinforcing inequities? What governance ensures equitable access and data sovereignty?	Participatory design, equity audits, technology assessment	Short-term (1–3 years)	All stakeholders	7, 8

(continued)

Table 12.1 (continued)

Priority area	*Key questions*	*Primary methods*	*Timeline*	*Stakeholders*	*Related chapters*
AI bias and resource use	What are effective AI applications? How to address bias in training data and energy/water footprints of AI infrastructure?	Algorithmic audits, case studies, life-cycle assessment	Short-term (1–3 years)	Researchers, practitioners, industry	8
Data interoperability	How to enable data sharing while protecting privacy and sovereignty? What standards support interoperability?	Technical assessments, governance-framework development	Medium-term (3–5 years)	Practitioners, policymakers, industry, researchers	8
Nature-based solutions business models	What are effective business models, financing mechanisms, and governance structures for nature-based solutions at scale?	Financial analysis, case studies, institutional assessment	Medium-term (3–5 years)	Practitioners, investors, industry, policymakers, researchers	9, 11
7. Health and well-being					
Climate-health nexus	What pathways link climate change to health outcomes?	Epidemiological studies, pathway analysis	Medium-term (3–5 years)	Researchers, practitioners	10
Mental health	What are psychological impacts and green-space benefits?	Psychological research, exposure studies	Medium-term (3–5 years)	Researchers, communities	9, 10
Environmental justice	How do interventions affect health inequalities?	Equity analysis, health-impact assessments	Ongoing priority	All stakeholders	7, 9, 10

Priority area	*Key questions*	*Primary methods*	*Timeline*	*Stakeholders*	*Related chapters*
8. Justice and equity					
Green-gentrification prevention	What policies effectively prevent displacement from climate and greening investments?	Comparative case studies, policy analysis, longitudinal monitoring	Short-term (1–3 years)	Policymakers, communities, practitioners	7, 10
Informal-settlement climate action	How can climate interventions support informal settlements without displacement? What co-design approaches work?	Participatory action research, co-design, case studies	Ongoing priority	Communities, practitioners, researchers	7
Energy poverty	What integrated approaches address energy poverty while advancing climate goals and reducing vulnerability?	Mixed methods, policy analysis, program evaluation	Short-term (1–3 years)	Policymakers, practitioners, communities	7, 10
Participatory budgeting	How does participatory budgeting improve climate-action equity and effectiveness? What design features enhance outcomes?	Comparative studies, impact assessment, participatory research	Medium-term (3–5 years)	All stakeholders	7, 11
Informal economy integration	How can informal workers and settlements be meaningfully included in climate planning and benefit from investments?	Participatory action research, policy analysis, co-design	Ongoing priority	Communities, investors, policymakers, practitioners, networks	7, 9
9. Biodiversity and ecology					
Urban biodiversity under climate change	How do climate impacts compound with other urban pressures (fragmentation, pollution, invasives) on biodiversity? What are tipping points?	Ecological monitoring, systems modeling, vulnerability assessments	Long-term (5+ years)	Researchers, practitioners, communities	9

Source Authors

- Networks: Transnational and regional city networks, e.g., UN-Habitat, C40, ICLEI, GCoM, UCLG
- All stakeholders: Coordinated action required across categories

Use Systems-Based Finance Tools (ICLEI TAP; World Bank APEX; EBRD Green Cities) and pursue green bonds with municipal capacity-building; engage with early-stage project development funds such as the Gap Fund, which supports more than 300 cities (Chapter 11).

Embed Justice Considerations Systematically: Procedural, distributive, and recognition justice dimensions should inform all stages of climate planning and implementation (Anguelovski et al., 2016; Chapter 7). This includes proactive measures addressing green gentrification risks (Bressane et al., 2024; Chapter 10), ensuring meaningful participation from marginalized communities, and incorporating diverse cultural perspectives into decision-making processes.

Use Data Strategically While Addressing Divides: Enhanced data-collection and monitoring systems track climate impacts, identify vulnerabilities, and evaluate interventions (Chapter 8). However, digital divides must be addressed to prevent excluding those most vulnerable from data-informed planning processes (Ricciardi & Callegari, 2022).

12.3.3 For Policymakers and Governance Bodies

Enable Multi-level Governance Systematically: The CHAMP's commitment to consulting and collaborating with subnational governments on climate action—including contributions to NDC 3.0—demonstrates promising institutional reforms (CPI & CCFLA, 2023; Chapter 11). National governments should explicitly recognize cities in climate strategies (UN-Habitat, 2025a, 2025b), devolve necessary authorities and resources, and establish coordinated investment planning that incorporates urban priorities. Urban sectors must be recognized as central for making the most of climate investment (CCFLA & SDSN Commission, 2025; Chapter 11), with city voices represented in governance structures.

Create Enabling Environments for Urban Climate Finance: Legal frameworks should clarify municipal borrowing authorities, debt-servicing capabilities, protections honoring commitments through political transitions, and project-execution powers (World Bank & UNCDF, 2024; Chapter 11). Intergovernmental fiscal transfers should be reliable,

Table 12.2 Overview of case studies from across the book

Case study	*Location*	*Focus area*	*Key innovation*	*Outcomes and impact*	*Chapter*
Paris 15-minute city	Paris, France	Polycentric urban form	Proximity-based development integrating land use, mobility, essential services within 15-minute walk/cycle	Reduced car dependency, enhanced service access, improved livability; ongoing equity monitoring to prevent displacement	10
Singapore Urban climate modeling	Singapore	AI-assisted climate analysis	Multi-scale microclimate models using AI to identify heat drivers, simulate scenarios, generate cooling recommendations	Evidence-based urban design interventions; informs building placement, materials, green infrastructure	8, 10
Stockholm Blue-Green infrastructure	Stockholm, Sweden	Ecosystem integration	Long-term integration of ecosystems, green spaces, nature-based solutions into resilient planning through 2030	Enhanced biodiversity, thermal regulation, ecosystem services; demonstrates policy commitment to nature-based approaches	9, 10
Amsterdam circular tendering	Amsterdam, the Netherlands	Circular-economy policy	Circular requirements for urban transformation, infrastructure renovation, demolition aligned with land allocation	Systemic shift toward resource efficiency; demonstrates regulatory approach at city scale; reduces virgin-material consumption	10

(continued)

Table 12.2 (continued)

Case study	*Location*	*Focus area*	*Key innovation*	*Outcomes and impact*	*Chapter*
CHAMP coalition	77 countries globally	Multi-level governance	National-subnational collaboration pledges in NDC 3.0; systematic consultation, unlocking climate finance	Formal city recognition in national strategies; potential to close ~40% of emissions gap between NDCs and 1.5 °C pathway	11
Brazil PCVR Program	Brazil	Climate federalism	Integration of urban, environmental, and climate policies across federal-state-municipal levels	Demonstrates vertical coordination; prioritizes metropolitan regions and vulnerable areas; strengthens multi-level cooperation	11
Denmark municipal climate Plans	Denmark	Multi-level governance	99% municipalities with climate plans; Climate Alliance coordination funded by Realdania and Local Government Denmark	Plans aim for 73% emissions cut by 2030 (exceeding 70% national target); 6500 + actions across sectors	11
Colombia just transition NDC	Colombia	Equity-centered planning	Multi-level NDC 3.0 consultations engaging cities, Indigenous/ Afro-Colombian groups, youth, women; WRI/GIZ support	Regional workshops identified adaptation priorities; emphasized peacebuilding and social inclusion as essential to climate action	7, 11

Case study	*Location*	*Focus area*	*Key innovation*	*Outcomes and impact*	*Chapter*
Alexandria Green City action plan	Alexandria, Egypt	Integrated climate planning	Comprehensive plan considering investment needs for adaptation/mitigation to 2030; World Bank capacity support	Significant adaptation investments (coastal protection, water); 50% greenhouse-gas savings potential by 2030 with $6.5B investment	11
Gap fund technical assistance	Accra, Ghana; Central Bosnia; Cape Town, South Africa	Project preparation	Technical assistance for bankable projects: waste-to-biogas (Accra), energy audits (Bosnia), stormwater plans (Cape Town)	Addresses early-stage capacity constraints; catalyzes investment by making projects "shovel-ready"; gender-responsive in Accra	11
Medellin Metrocable	Medellin, Colombia	Accessible mobility	Cable-car system enhancing territorial accessibility for low-income periphery neighborhoods; transit-network integration	Reduced mobility exclusion; improved access to employment, education, services for marginalized communities	7, 10
Lima air quality monitoring	Lima, Peru	Data-driven intervention	Expanded air-quality monitoring capacity by 176%; five modules focused on children's pollution exposure; network support	Granular data enables targeted interventions; demonstrates networks facilitating rapid dissemination; informs policy/awareness	5, 8

(continued)

Table 12.2 (continued)

Case study	*Location*	*Focus area*	*Key innovation*	*Outcomes and impact*	*Chapter*
Freetown tree campaign	Freetown, Sierra Leone	Nature-based solutions	Tree-planting campaign digitally tracked over 550,000 trees via mobile technology; created more than 600 green jobs; network support	Combines mitigation/adaptation with job creation and engagement; demonstrates the feasibility of nature-based solutions in resource-constrained contexts	5, 9, 10
Quezon City GrowQC	Quezon City, Philippines	Urban agriculture	Transformed idle land into productive farms addressing food security; 4100 urban farmers engaged; community participation	Demonstrates food security–climate adaptation–empowerment nexus; idle-land activation provides low-cost pathway	7, 10
Barcelona energy advice points	Barcelona, Spain	Energy poverty, behavior change	Energy advice combining practical support for 222,000 vulnerable residents with climate education, building community capacity	Integrates immediate needs with long-term goals; equity-centered approach	7
Indigenous-municipal partnerships	Various Canadian cities	Co-design processes	Long-term relationship-building, reciprocal learning, power-sharing arrangements; Yarning about Urban Country approach	Contextually appropriate responses incorporating indigenous knowledge; generates solutions unattainable through conventional planning	6, 9

Case study	*Location*	*Focus area*	*Key innovation*	*Outcomes and impact*	*Chapter*
Disinformation response playbook	C40 cities and ICLEI Canada	Disinformation response	First playbook co-created by researchers, city representatives, experts from civil society, tech sector, media	Provides cities with tools to counter climate disinformation; strengthens resilience against attacks on climate action	6, 8
CityStudio Vancouver	Vancouver, Canada; Athens, Greece	University-city partnerships	Civic-engagement lab connecting municipal government, universities, community partners to co-create climate solutions	Opportunities for thousands to collaborate on more than 2600 projects through growing global network	6
Nairobi informal settlement upgrading	Nairobi, Kenya	Procedural justice	Community-led planning integrating climate considerations in informal settlements; participatory vulnerability assessments	Enhanced resilience for marginalized populations; demonstrates importance of procedural justice	7
Medellin social urbanism	Medellin, Colombia	Integrated development	Libraries, cable cars, public spaces in low-income areas; integrating infrastructure, social services, economic opportunity	Reduced inequalities while building climate resilience; demonstrates integrated approach	7, 10
Los Angeles LADWP just transition	Los Angeles, USA	Just-transition strategies	Community-based renewable energy strategies and equitable transition planning with Just Transition task force	Direct dialogue with workers and frontline communities; moving from strategy to implementation plan	7, 10

(continued)

Table 12.2 (continued)

Case study	*Location*	*Focus area*	*Key innovation*	*Outcomes and impact*	*Chapter*
Dark matter labs	Copenhagen, Denmark	Systems innovation	Transdisciplinary collective focused on systems-level design, innovation, experimental civic infrastructure	Horizontal boundary-spanning; creates institutional innovations for urban transformation	4
OCULUS studio	Australia	Landscape regeneration	Climate-resilient urban regeneration through landscape architecture, policy influence, multidisciplinary collaboration	Vertical boundary-spanning; demonstrates practitioner-led climate action	4
Menjez local council	Menjez, Lebanon	Regional development	Regional municipality as climate model, building multi-sector collaborations, getting the most benefits from transborder projects	Spatial and horizontal boundary-spanning; demonstrates small-city leadership	4
Brazil PAC Program (NOVO PAC)	Brazil	Multi-level governance, infrastructure	US$1.7 trillion budget integrating federal-state-city collaboration on infrastructure; explicit multi-level framework	Demonstrates large-scale, multi-level climate investment; prioritizes vulnerable areas; includes urban climate resilience	11
Melbourne 20-minute neighborhoods	Melbourne, Australia	Proximity planning	Whole-of-government approach embedding 20-minute neighborhoods in infrastructure plans; 2017–2050 framework	Mainstreams accessibility and climate goals; demonstrates long-term planning commitment	10

Case study	*Location*	*Focus area*	*Key innovation*	*Outcomes and impact*	*Chapter*
City creditworthiness initiative	Global south cities (2013–2023)	Municipal capacity-building	World Bank program: Creditworthy Academies, multi-year technical assistance, mentoring of graduates	Strengthened municipal capacity to access capital markets; created cohort of trained officials	11
Green & sustainability Bonds – ASEAN	Malaysia, Singapore, Indonesia	Municipal climate finance	Policy frameworks and tax incentives; Mexico City first Global South bond (US$50M)	US$6.2 trillion issued since 2010; demonstrates viable municipal-financing pathway	11
ICLEI TAP	Global (50+ countries, 160+ projects)	Finance access	Platform supporting locally led sustainability-project financing; technical support for project development	Facilitated financing for diverse climate projects; demonstrates role of intermediary platforms	11
Climate view platform	Multiple cities globally	Digital climate planning	Integrated digital platform for climate-action planning, tracking, and decision support	Enables evidence-based climate planning; facilitates progress tracking	8
Brazil CHAMP – CityCatalyst	Brazil	Digital infrastructure for NDCs	Digital stack for subnational climate-action tracking and reporting; enables cities to demonstrate contributions	Transparent subnational climate data; supports NDC implementation	8, 11
Seoul community service centers	Seoul, South Korea	Integrated public services	Reducing institutional isolation through integrated facilities offering education, culture, health, recreation	Enhanced administrative coherence; improved public trust; demonstrates social infrastructure approach	10

(continued)

Table 12.2 (continued)

Case study	*Location*	*Focus area*	*Key innovation*	*Outcomes and impact*	*Chapter*
Medellin UVAs	Medellin, Colombia	Social infrastructure	Interdependent facilities offering education, culture, health, and recreation under one roof in underserved areas	Strengthened community cohesion; improved service access in marginalized areas	7, 10
Kampala Kisenyi informal settlement	Kampala, Uganda	Informal-settlement adaptation	Community-based climate action in settlements housing 60% of metropolitan residents; participatory planning	Demonstrates feasibility of climate action in informal contexts; builds community capacity	7
Accra informal workers integration	Accra, Ghana	Informal economy, justice	Participatory approaches incorporating informal workers' experiences into climate-intervention design	Ensures that climate action includes informal economy; reduces displacement risk	7
Stockholm environment programme 2030	Stockholm, Sweden	Environmental planning	Integrated environmental planning addressing climate, biodiversity, and pollution through 2030	Demonstrates policy integration across environmental domains; long-term sustainability commitment	9, 10, 11
Mexico City environmental policy	Mexico City, Mexico	Urban environmental governance	Comprehensive evaluation of environmental policy 2018–2022; systematic assessment of interventions	Demonstrates importance of policy evaluation; identifies successes and gaps; informs adaptive management	10, 11

Source Authors

adequate, and aligned with climate objectives. Regulatory frameworks supporting innovative financing instruments require development. The capacity for multilateral financing institutions to deal with municipal-level actors and vice versa should be built in alignment with sovereign state-level coordination mechanisms.

Integrate Urban Content in National Climate Commitments: Analysis of 194 NDCs (June 2023) shows 28% indicate cities as priority and 39% mention cities (Tollin et al., 2024; Chapter 11). Updated analysis of NDCs 3.0 submitted by 2025 shows a marked increase in urban integration, with the number of NDCs containing strong urban content *nearly doubling* compared with previous submissions (UN-Habitat, 2025a, 2025b). Approximately *three-quarters* of NDCs now reference cities or local governments within their mitigation and adaptation strategies, while *two-thirds* explicitly reference multilevel governance mechanisms (UN-Habitat, 2025a, 2025b). Urban sector responses also intensified, with *nearly half* of NDCs including actions related to housing, informal settlements, or other key urban sectors (UN-Habitat, 2025a, 2025b; WRI, 2025).

These updated findings reveal progress and significant further improvement opportunities for NDCs 3.0, including: (1) establishing robust urban greenhouse gas emissions and climate-risk baselines, (2) identifying urban challenges and corresponding response measures, and (3) detailing actions with their associated means of implementation, including finance, technology transfer and capacity-building (Ranalder et al., 2025). Future submissions should systematically incorporate city contributions, recognizing that 75% of GCoM signatories already exceed national commitments (GCoM, 2024). Without necessary capacities and resources, commitments risk remaining aspirational. Cities' demonstrated capacity for ambitious climate action illustrates substantial untapped potential for accelerated implementation across NDCs and other national climate-policy processes.

Support regional and City-Specific Research and Action: Funding mechanisms should prioritize research addressing diverse urban contexts, particularly rapidly urbanizing regions of the Global South where climate vulnerabilities concentrate (Bai et al., 2018; UN DESA, 2018; UNESCAP, 2020; Chapter 10). This includes supporting data-collection infrastructure, capacity-building for local researchers and community-engaged research, and mechanisms translating global knowledge for local applications.

Advance Culture-Climate Integration: National and local frameworks should facilitate integration of cultural heritage, Indigenous Knowledge systems, and multispecies justice considerations into climate planning (Ives & Baker, 2024; Chapter 9). This includes supporting community-engagement processes that recognize fluid cultural identities (Gulsrud et al., 2018) and challenge business-as-usual power imbalances constraining transformative action (Frantzeskaki et al., 2024; Chapter 10). Crucial to this integration is not only diversifying who is in the room, but how those in the room are engaged; this is critical to rectify power imbalances and encourage meaningful participation.

Establish Legal Frameworks clarifying municipal borrowing authorities and create enabling environments for innovative financing instruments; support country platforms for climate action that integrate urban priorities as central for gaining the most benefit from climate investment (Chapter 11).

12.3.4 Cross-Cutting Imperatives

Accelerate Implementation Speed and Scale: Current climate change action is insufficient to achieve the goals of the Paris Agreement (UNFCCC, 2025). All stakeholders must prioritize rapid scaling of proven solutions while continuing to innovate and learn from failure. This requires overcoming risk aversion, streamlining approval processes, and accepting adaptive management approaches that acknowledge uncertainty (Folke et al., 2021).

Strengthen Science–Policy–Practice Interfaces: The boundary-spanning mechanisms and sustained collaboration exemplified by Innovate4Cities should be replicated across domains (Chapter 4). Regular convenings that facilitate dialogue, shared digital platforms that support knowledge exchange, and institutional mechanisms that embed ongoing collaboration are proving essential (Hale, 2016; Hickmann et al., 2021).

Address the "implementation gap" Systematically: While many urban climate solutions are well-understood (Seto et al., 2014), catalyzing widespread implementation—particularly in small and mid-sized cities—remains challenging. Targeted technical-assistance programs (e.g.,—the World Bank's APEX city tool and the EBRD's Green Cities program) (Chapter 11) and sector-specific readiness-assessment tools that allow cities to understand financing options available to them based on their financial and governance fitness are central to bridging this gap.

Communicate Across Multiple Languages: Recognizing co-benefits and socioeconomic factors and employing contextual communication strategies enable climate-forward opportunities to enter the mainstream. This includes translating technical concepts across cultures, languages, and professional domains, avoiding jargon, and emphasizing immediate local benefits alongside global climate objectives (Yeh et al., 2025; Fig. 12.1).

12.4 Critical Research Priorities

Analysis across chapters reveals interconnected research priorities informing future Innovate4Cities initiatives:

12.4.1 Understanding Climate Overshoot Implications for Cities

Emerging projections suggest that exceeding the 1.5 °C Paris Agreement limit is highly probable, with potential for "overshoot" scenarios (which include exceeding and returning to a particular temperature goal) carrying potentially deleterious and irreversible effects for urban infrastructure, ecosystems, and people, particularly for vulnerable populations (Gasser et al., 2015). Research priorities include:

- **Compound and Cascading Risk Assessment**: Understanding how climate hazards interact with other hazards (natural, human-caused, biological), existing urban vulnerabilities, infrastructure interdependencies, and socioeconomic inequalities in different regions to generate compound risks exceeding the sum of individual impacts (Kapucu et al., 2024). Understanding the risks arising from mitigation responses to effectively achieve a decline in global temperature (Reisinger et al., 2025).
- **Adaptation Limits and Transformational Responses**: Identifying thresholds beyond which incremental adaptation proves insufficient, requiring transformational change in urban form, governance, or function (Lwasa et al., 2022). Identifying the potential irreversible harm that cities in different regions may experience (Reisinger et al., 2025), including the assessment of loss and damage.
- **Regenerative City Concepts**: Advancing resilience, as an evolutionary and transformative process, encompassing regenerative approaches, enabling cities to restore ecological functions,

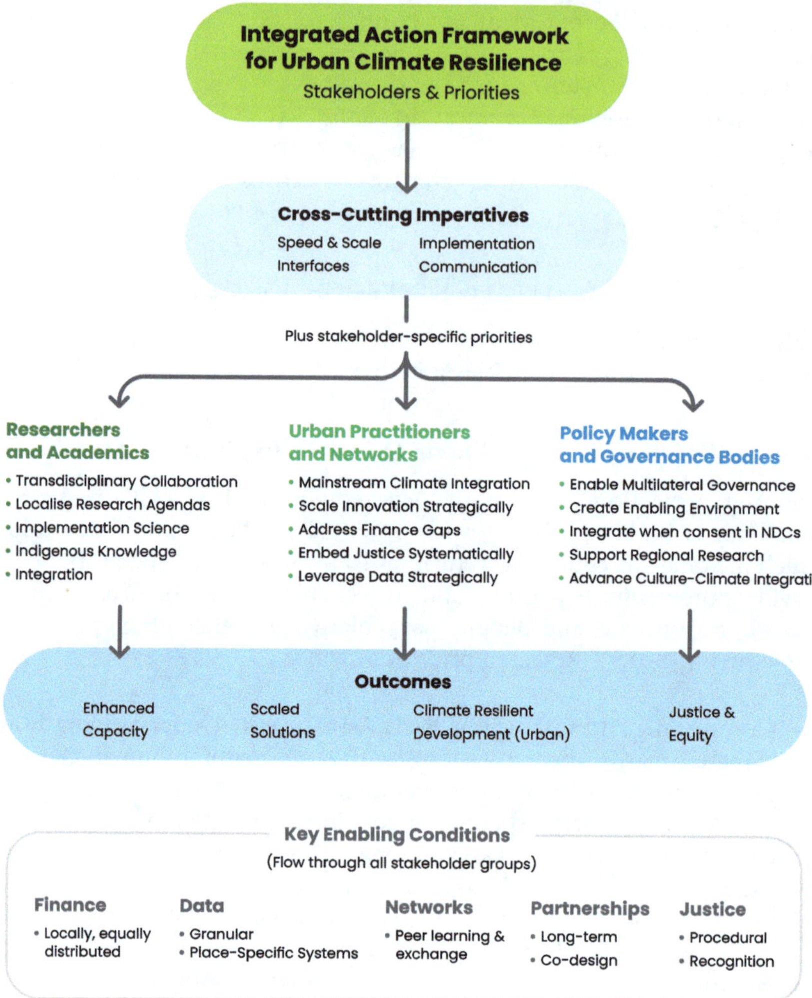

Fig. 12.1 Integrated action framework for urban climate resilience (*Source* Authors, Image created by Genov, N.)

enhance biodiversity, and contribute positively to planetary health (Chapter 10), as well as reducing vulnerability and exposure for people and infrastructures.

12.4.2 Localizing Knowledge and Research

Persistent gaps in geographically diverse, context-specific research require prioritization (Chapter 9):

- **Global South Urban Climate Science**: Expanding research coverage that addresses rapidly urbanizing regions where climate vulnerabilities concentrate and most population growth will occur through 2050 (UN-Habitat, 2024).
- **Small and Mid-Sized City Research**: Moving beyond a dominant focus on megacities toward understanding challenges and opportunities in smaller urban centers that lack extensive technical and financial resources, yet collectively house substantial populations (Kendal et al., 2020). Including data and analysis at the level of small-scale neighborhoods to address inequity and disproportionate risk.
- **Integration of Indigenous Knowledge and Local Knowledge**: Developing robust methodologies for systematic incorporation of diverse knowledge systems into urban climate research and planning, moving beyond tokenistic consultation (Cumpston et al., 2022; Tengö et al., 2021; Chapters 6, 10).
- **Culturally Appropriate Indicators**: Identifying measurement frameworks reflecting diverse community priorities and values, rather than imposing standardized metrics potentially misrepresenting local contexts. Informal systems within urban settlement must be recognized for meeting local needs absent formal structures (Chapter 10).
- **Regional-level Research**: Moving beyond the global discussion on climate change to better understand the regional differences in climate change impacts and responses in urban areas.

12.4.3 Advancing Implementation Science

Understanding how to implement and accelerate solutions (Chapter 2) requires:

- **Scaling Mechanisms and Transferability**: Examining conditions enabling successful adaptation and replication of innovations across diverse contexts, including institutional, financial, technical, and cultural factors (Collier et al., 2023; Chapter 5).
- **Municipal Capacity-Building Pathways**: Identifying effective approaches for strengthening technical, financial, and governance capacities, particularly in under-resourced municipalities (Hurlimann et al., 2020; Chapter 11).
- **Behavioral Change and Social Innovation**: Understanding drivers of individual and collective behavior changes necessary for sustainable consumption patterns, mobility shifts, and support for transformative policies (Chapter 10).
- **Political Economy of Urban Climate Action**: Analyzing power dynamics, vested interests, and political-feasibility considerations shaping implementation prospects, focusing on lessons to be learned from informal urban settlements (Chapter 7) and non-codified practices and behaviors.

12.4.4 Finance and Economic Dimensions

Critical research gaps persist (Chapters 9–11) regarding:

- **Investment Needs and Financing Mechanisms**: Refining estimates of capital requirements across sectors, city typologies, and development contexts, while identifying effective mechanisms that mobilize diverse capital sources (Deuskar et al., 2025).
- **Economic Impacts of Climate Action and Inaction**: Quantifying costs of delayed action, co-benefits of early intervention, and analyzing distributional impacts of different policy approaches.
- **Circular-Economy Pathways**: Understanding economic, social, and environmental implications of transitioning from linear to circular resource flows, including employment impacts and business-model innovations (Ruini et al., 2025).

- **Municipal Creditworthiness Determinants**: Identifying factors that enhance cities' access to capital markets and innovative financing instruments, particularly in developing-country contexts.

12.4.5 *Governance and Institutional Dimensions*

Advancing effective urban-climate governance requires research on:

- **Multi-level Governance Effectiveness**: Evaluating mechanisms that coordinate action across governance levels (UN DESA, 2024), identifying barriers to vertical and horizontal integration, and assessing role of city networks and non-state actors (Chapter 5).
- **Policy Integration and Mainstreaming**: Understanding how climate considerations can be systematically embedded across sectoral policies (Adams et al., 2023) and isolated and disconnected departments while maintaining coherence.
- **Participatory Governance and Co-production**: Examining conditions enabling meaningful community engagement, power-sharing arrangements, and co-production of knowledge and solutions, and identifying informal settlement lessons (Almansi et al., 2011; Chapters 6 & 7).
- **Adaptive Governance Under Uncertainty**: Developing frameworks supporting flexible, learning-oriented approaches that accommodate irreducible uncertainties while maintaining accountability.

12.4.6 *Technology and Innovation*

Critical questions regarding technological dimensions (Chapter 8) include:

- **AI and Digitalization Implications**: Assessing opportunities and risks associated with artificial intelligence and digital technologies for urban climate action, including equity considerations, energy footprints (Simon & O'Brien, 2023), and governance requirements.
- **Nature-based Solutions and Ecosystem Based Adaptation Effectiveness**: Strengthening the evidence base regarding ecosystem service provision, biodiversity co-benefits (Langemeyer & Baró,

2021; Chapter 10), long-term performance under climate change (Reynolds et al., 2020), and optimal design parameters.
- **Infrastructure Resilience and Adaptation**: Understanding how gray, green, and hybrid infrastructure systems can be designed and retrofitted for enhanced resilience to intensifying climate hazards (Rosenzweig et al., 2018; Chapter 10).

12.4.7 Health and Well-Being Dimensions

Recognizing urban health as both climate impact and co-benefit (Chapter 10). Distinguishing physical health impacts between the environmental and the habitual (for example, communicable and non-communicable) is crucial:

- **Climate-Health Nexus**: Deepening understanding of pathways through which climate change affects urban health outcomes and how climate action generates health co-benefits (WHO, 2025).
- **Mental Health and Well-Being**: Examining psychological dimensions of climate change and green-space exposure (Houlden et al., 2018; Chapter 10), moving beyond physical health outcomes.
- **Environmental Justice and Health Equity**: Understanding how climate hazards and adaptation interventions affect health inequalities across socioeconomic and demographic groups (Kabisch et al., 2023; Chapter 7).

Table 12.1 presents a matrix of research priorities, synthesized through keyword analysis from across the book chapters, to guide research-funding decisions, university curriculum development, practitioner-training programs, and collaborative-research agenda-setting across the Innovate4Cities network and broader urban-climate community.

12.5 Next Steps in the Roadmap

12.5.1 The Systems Imperative

Synthesis across chapters reveals that addressing urban climate challenges requires transcending the disconnections among sectors toward integrated systems thinking. Cities function as complex social–ecological–technological systems (SETS) where infrastructures, institutions, and practices dynamically interact (Grimm et al., 2015; McPhearson et al., 2016; Chapter 4). Nexus conceptualizations progress beyond fragmented interventions toward holistic frameworks addressing polycrisis challenges; that is, the interlinked, compounding threats posed by biodiversity loss, climate change, inequity, and injustice (Irvin et al., 2024; Chapter 10).

The GRAA's evolution from concentric wheel to building structure exemplifies a maturing capacity to represent system complexity without overwhelming stakeholders (Irvin et al., 2024; Prieur-Richard et al., 2019; Chapter 3). Four interconnected transformation realms emerge:

Land and Ecosystems (Chapter 9): Polycentric urban forms integrating proximity principles with ecosystem connectivity through green–blue infrastructure networks support biodiversity while providing thermal regulation, stormwater management, and social benefits (Langemeyer & Baró, 2021; Moreno et al., 2021; Van Nes, 2021).

Physical and Social Infrastructure and Public Services (Chapter 10): Infrastructure can be reconceptualized to encompass social and civic structures alongside technical systems, recognizing community hubs, libraries, and care centers as essential resilience foundations (Hollis et al., 2023; Renner et al., 2024). Seoul's Community Service Centers and Medellín's Unidades de Vida Articulada exemplify interdependent facilities offering education, culture, health, and recreation under one roof, enhancing administrative coherence and public trust.

Resource Flows and Circular Economy (Chapter 10): Material loops can be closed through reduction, reuse, repair, refurbishment, remanufacturing, recycling, and recovery, supported by regulatory frameworks, fiscal tools, and systematic metabolic modeling (UNEP-UN-Habitat, 2021; Williams, 2022). Cities account for over 75% of global material-resource consumption (IRP-UNEP, 2018), with construction sectors remaining among the most resource-intensive urban-system components (Li et al., 2019).

Governance and Social Structures (Chapter 11): Multi-level partnerships that align local action with national commitments (UN

DESA, 2024), participatory processes ensure inclusive decision-making (Anguelovski et al., 2016; Chapter 7), and capacity-building, enabling effective implementation. The CHAMP coalition's 77 member countries demonstrate formal recognition that systematic collaboration between governance levels maximizes climate impact (ICLEI, 2024).

12.5.2 *Catalyzing Transformative Change*

Our review across this book's chapters identifies four critical catalysts for accelerating transformation:

Integrated Frameworks: Tools such as SETS, adaptive pathways, and the Nature-Futures-Framework (Pereira et al., 2020) systematically map innovations, identify progressive-action pathways, and serve as organizing structures for transdisciplinary co-production. These frameworks help navigate complexity rather than removing it, providing analytical scaffolding for understanding trade-offs and synergies across interventions (Chapter 4).

Narrative Shifts: Progressing toward plural, decolonial understandings of innovation processes acknowledges and respects local communities—including vulnerable, marginalized groups from intersectional justice perspectives—in setting platforms for dialogue between knowledge holders (Amorim-Maia et al., 2022; Frantzeskaki et al., 2024; Chapters 7, 9). Language can connect as well as divide; plurality and inclusivity may enhance innovations with nature in cities.

Cultural Integration: Embedding cultural considerations—including heritage, Indigenous and Local Knowledge, and multispecies justice frameworks (Celermajer et al., 2021; Chapter 9)—into development, planning, and governance can contribute to simultaneously addressing biodiversity loss, climate change, and inequality. Colonial processes of urbanization contributed to dispossession and habitat destruction; transformative approaches embedding cultural considerations and Traditional knowledges may more effectively challenge business-as-usual power imbalances (Gazing Wolf et al., 2024; Chapters 6, 9).

Forward-Looking Learning: Employing learning to inform future actions through dialogues between researchers, practitioners, and communities can help integrate citizen science with structured data collection and diverse knowledge systems (Tengö et al., 2021; Chapters 8, 9).

Careful selection of indicators that effectively represent ecosystem structure and function while allowing time and place comparisons supports both local monitoring efforts and global reporting processes.

12.5.3 Persistent Challenges

Despite progress documented throughout the book, significant challenges remain:

Finance Gaps (Chapter 11): Current flows represent only 16–33% of requirements (Buchner et al., 2023; Lwasa et al., 2022), with particularly acute deficits in Global South cities lacking creditworthiness, technical capacity, or national-government support for accessing innovative financing mechanisms (Deuskar et al., 2025). The 2024 State of Cities Climate Finance report reveals that while annual urban climate finance during 2021–2022 reached US$830 billion, this falls significantly below the US$4.5 trillion needed by 2030 (CCFLA, 2024).

Political-Economy Barriers: Electoral cycles driving short-term focus, vested interests resisting transformative change, and risk aversion constraining innovation adoption slow transformation despite mounting urgency (Bulkeley & Newell, 2023; Chapter 11). Current and planned adaptation measures cannot resolve most urban-climate risks, creating persistent adaptation gaps across all regions and hazard types (Lwasa et al., 2022).

Capacity Constraints: Many municipalities—particularly small and mid-sized cities—lack technical expertise, financial resources, or institutional capabilities for effective climate-action planning and implementation (Hurlimann et al., 2020; Chapter 9). Climate change literacy among public officials varies significantly, with knowledge gaps directly affecting policy-implementation capacity (Yeh et al., 2025; Chapter 8).

Knowledge Gaps (Chapters 9–11): While expanding, urban climate science remains geographically skewed toward Global North cities (Kendal et al., 2020), with insufficient research addressing diverse contexts and the integration of Indigenous and Local Knowledge (Orlove et al., 2023; Chapter 6), and implementation pathways. The "city-size bias" in knowledge production means that rapidly urbanizing small and mid-sized cities receive insufficient research attention relative to their aggregate importance.

Insufficient Scientific Engagement (Chapter 2): To ensure progress at the science–policy–practice nexus, scientists need to be adequately engaged throughout the Innovate4Cities process. From conferences and marketplaces to knowledge-product developments, ensuring that the scientific foundation is strengthened represents an ongoing challenge and opportunity to tackle intensifying climate risks, hazards, and overshoot scenarios (Solecki et al., 2021).

12.5.4 The Path Forward

The GRAA provides a robust framework for organizing research and innovation priorities, actions, and gaps, alongside established science–practice networks offering mechanisms for sustained collaboration, balancing accessibility and coherence, although depth, and diversity of engagement remain areas requiring attention (Hunter et al., 2023; Irvin et al., 2024; Oke et al., 2022; Chapters 2, 5). Hybrid conference formats demonstrate potential for inclusive participation while maintaining benefits of face-to-face collaboration (Walsh et al., 2022). Dedicated funding for Global South participation, regional preparatory meetings, and sophisticated hybrid technologies should be considered to further enhance inclusive engagement.

The processes associated with author selection and sector engagement for the SRCCC demonstrated deliberate IPCC leadership in inviting practitioners' expertise for reality checks on regionally relevant solutions that reflect implementation feasibility and effectiveness alongside scientific rigor. Coordination among practitioners, partly enabled through Innovate4Cities, facilitates structured inputs during scoping and expert-review processes, establishing mechanisms for continued science–policy–practice collaboration as the global urban climate community advances toward the 2027 SRCCC release.

12.6 Conclusion

This book documents a six-year journey bridging traditionally isolated communities to address urban climate challenges. From the 2016 proposal for a Special Report by South Africa (IPCC, 2016) through three GRAA iterations and expanding participation, Innovate4Cities demonstrates cities' transformative potential when diverse actors commit to shared learning and action.

Three critical insights emerge:

First, integrating knowledge and innovation requires collaboration between academic and practitioner perspectives, combining rigorous research with real-world implementation insights.
Second, solution implementation and scaling represent the critical next phase. While many urban climate solutions are well-understood, catalyzing widespread implementation—particularly in small and mid-sized cities—has become paramount.
Third, communication across disciplinary, cultural, and professional languages determines success, enabling mainstreaming of local climate solutions through recognition of co-benefits and contextual strategies.

As cities confront intensifying climate hazards while contributing 67–72% of global emissions, ambitious mitigation and transformative adaptation demand urgent response. Current trajectories prove insufficient: climate finance gaps persist, inequality widens, and exceeding 1.5 °C appears increasingly likely, necessitating preparation for overshoot scenarios while accelerating emissions reductions.

Yet grounds for optimism exist. Through Innovate4Cities, cities have demonstrated capacity for ambitious action, while governments increasingly commit to multi-level partnerships. The GRAA shows a path forward through strengthening collaborative frameworks, accelerating implementation through capacity-building and innovative financing, and ensuring that change advances justice and equity—particularly for the 3 billion people projected to live in informal settlements by 2050, whose community-led innovations must be integrated alongside formal urban systems.

As IPCC authors prepare the AR7 SRCCC (due 2027), the knowledge foundations, collaborative networks, and implementation insights generated through Innovate4Cities provide essential inputs. This journey demonstrates that meaningful progress is achievable through sustained collaboration, adaptive learning, and shared purpose.

The window for action narrows, but the tools, knowledge, and collaborative frameworks now exist to chart a more resilient, equitable, and just urban future. The question is no longer whether cities can lead climate action, but whether we will seize this opportunity with the urgency it demands.

The work continues—and it must.

Acknowledgements The authors acknowledge Chiara Morfeo, of C40 Cities, for her contributions to the development of this chapter.

Competing Interests

The authors have no conflicts of interest to declare.

References

Acuto, M., & Leffel, B. (2020). Understanding the global ecosystem of city networks. *Urban Studies, 58*(9), 1758–1774. https://doi.org/10.1177/0042098020929261

Adams, C., Frantzeskaki, N., & Moglia, M. (2023). Mainstreaming nature-based solutions in cities: A systematic literature review and a proposal for facilitating urban transitions. *Land Use Policy, 130*, Article 106661. https://doi.org/10.1016/j.landusepol.2023.106661

Almansi, F., Hardoy, A., Hardoy, J., Pandiella, G., Tambussi, L., Urquiza, G., McGranahan, G., & Satterthwaite, D. (2011). *Limits to participation: The struggle for environmental improvement in Moreno, Argentina*. IIED.

Amorim-Maia, A. T., Anguelovski, I., Chu, E., & Connolly, J. (2022). Intersectional climate justice: A conceptual pathway for bridging adaptation planning, transformative action, and social equity. *Urban Climate, 41*, Article 101053. https://doi.org/10.1016/j.uclim.2021.101053

Anguelovski, I., & Connolly, J. J. (2022). The green city and social injustice: 21 Tales from North America and Europe. *Routledge*. https://doi.org/10.4324/9781003183273

Anguelovski, I., Shi, L., Chu, E., Gallagher, D., Goh, K., Lamb, Z., Reeve, K., & Teicher, H. (2016). Equity impacts of urban land use planning for climate adaptation: Critical perspectives from the global north and south. *Journal of Planning Education and Research, 36*(3), 333–348.

Bai, X., Dawson, R. J., Ürge-Vorsatz, D., Delgado-Ramos, G. C., Barau, A. S., Dhakal, S., Dodman, D., Leonardsen, L., Masson-Delmotte, V., Roberts, D., & Schultz, S. (2018). Six research priorities for cities and climate change. *Nature, 555*(7694), 23–25. https://doi.org/10.1038/d41586-018-02409-z

Bartlett, C., Marshall, M., & Marshall, A. (2012). Two-eyed seeing and other lessons learned within a co-learning journey of bringing together indigenous and mainstream knowledges and ways of knowing. *Journal of Environmental Studies and Sciences, 2*(4), 331–340. https://doi.org/10.1007/s13412-012-0086-8

Bevacqua, E., Schleussner, C. F., & Zscheischler, J. (2025). A year above 1.5 °C signals that Earth is most probably within the 20-year period that will reach the Paris Agreement limit. *Nature Climate Change, 15*, 262–265. https://doi.org/10.1038/s41558-025-02246-9

Bressane, A., Cunha Pinto, J. P., & Castro Medeiros, L. C. (2024). Countering the effects of urban green gentrification through nature-based solutions: A scoping review. *Nature-Based Solutions, 5*, Article 100131. https://doi.org/10.1016/j.nbsj.2024.100131

Buchner, B., Naran, B., Padmanabhi, R., Stout, S., Strinati, C., Wignarajah, D., Miao, G., Connolly, J., & Marini, N. (2023). *Global landscape of climate finance 2023.* Climate Policy Initiative. https://www.climatepolicyinitiative.org/publication/global-landscape-of-climate-finance-2023/

Bulkeley, H., & Betsill, M. (2005). Rethinking sustainable cities: Multilevel governance and the 'urban' politics of climate change. *Environmental Politics, 14*(1), 42–63. https://doi.org/10.1080/0964401042000310178

Bulkeley, H., & Betsill, M. (2013). Revisiting the urban politics of climate change. *Environmental Politics, 22*(1), 136–154.

Bulkeley, H., & Newell, P. (2023). *Governing climate change* (3rd ed.). Routledge. https://doi.org/10.4324/9781003219828

Bush, J., West, K., & Miller, M. (2023). Yarning about Urban Country: Knowing, learning and planning for healthy, liveable cities. In *Urban Country* (pp. 1–24). Routledge. https://doi.org/10.4324/9781003108757-3

C40 Cities and Global Covenant of Mayors. (2025). *From Paris to Belém: A decade of local climate leadership.* https://www.globalcovenantofmayors.org/wp-content/uploads/2025/06/From-Paris-to-Belem-A-Decade-of-Local-Climate-Leadership-.pdf

Cannon, A. J. (2025). Twelve months at 1.5 °C signals earlier than expected breach of Paris Agreement threshold. *Nature Climate Change, 15*, 266–269. https://doi.org/10.1038/s41558-025-02247-8

Castree, N., Barnes, T., & Salmond, J. (Eds.). (2025). *Making geography matter: The past and present of a changing discipline.* Routledge.

CCFLA. (2024). *The state of cities climate finance 2024.* Cities Climate Finance Leadership Alliance. https://citiesclimatefinance.org/publications/2024-state-of-cities-climate-finance

CCFLA and SDSN Commission. (2025). *Integrating urban and subnational priorities into country platforms: Policy brief.* https://citiesclimatefinance.org/publications/integrating-urban-and-subnational-priorities-into-country-platforms

Celermajer, D., Schlosberg, D., Rickards, L., Stewart-Harawira, M., Thaler, M., Tschakert, P., Verlie, B., & Winter, C. (2021). Multispecies justice: Theories, challenges, and a research agenda for environmental politics. *Environmental Politics, 30*(1–2), 119–140. https://doi.org/10.1080/09644016.2020.1827608

City of Freetown. (2023). #FreetownTheTreeTown Campaign Report. https://www.freetown.gov.sl/

City of Stockholm. (2024). *Environment Programme 2030*. https://start.stockholm/globalassets/start/om-stockholms-stad/sa-arbetar-staden/miljo-och-klimat/environment-programme-2030.pdf

Collier, M. J., Frantzeskaki, N., Connop, S., Dick, G., Dumitru, A., Dziubała, A., Fletcher, I., Georgiou, P., Hölscher, K., Kooijman, E., Lodder, M., Madajczyk, N., McQuaid, S., Nash, C., Osipiuk, A., Quartier, M., Reil, A., Rhodes, M.-L., Rizzi, D., …, & Xidous, D. (2023). An integrated process for planning, delivery, and stewardship of urban nature-based solutions: The connecting nature framework. *Nature-Based Solutions, 3*, Article 100060. https://doi.org/10.1016/j.nbsj.2023.100060

CPI & CCFLA. (2023). *Financial aggregation blueprints for urban climate infrastructure—Annex 2: Municipal green bonds blueprints*. https://www.climatepolicyinitiative.org/wp-content/uploads/2023/06/Annex-2-Municipal-Green-Bonds-Blueprints.pdf

Creutzig, F., Lohrey, S., Bai, X., Baklanov, A., Dawson, R., Dhakal, S., Lamb, W. F., McPhearson, T., Minx, J., Munoz, E., & Walsh, B. (2019). Upscaling urban data science for global climate solutions. *Global Sustainability, 2*, Article e2. https://doi.org/10.1017/sus.2018.16

Cucca, R., Friesenecker, M., & Thaler, T. (2023). Green gentrification, social justice, and climate change in the literature: Conceptual origins and future directions. *Urban Planning*, *18*(1), 283–295. https://doi.org/10.17645/up.v8i1.6129

Cumpston, Z., Fletcher, M.-S., & Head, L. (2022). *Plants: Past, present and future*. Thames and Hudson.

Deuskar, C., Murray, S., Leiva Molano, J. S., Khan, I. A., & Maria, A. (2025). *Banking on cities: Investing in resilient and low-carbon urbanization*. Urban Development Series. World Bank. https://doi.org/10.1596/978-1-4648-2222-3

Elmqvist, T., Fragkias, M., Goodness, J., Güneralp, B., Marcotullio, P. J., McDonald, R. I., Parnell, S., Schewenius, M., Sendstad, M., Seto, K. C., & Wilkinson, C. (2013). Urbanization, biodiversity and ecosystem services: Challenges and opportunities. *Springer*. https://doi.org/10.1007/978-94-007-7088-1

Folke, C., Polasky, S., Rockström, J., Galaz, V., Westley, F., Lamont, M., Scheffer, M., Österblom, H., Carpenter, S. R., Stuart Chapin III, F., Seto, K. C., Weber, E. U., Crona, B. I., Daily, G. C., Dasgupta, P., Gaffney, O., Gordon, L. J., Hoff, H., Levin, S. A., …, & Walker, B. H. (2021). Our future in the Anthropocene biosphere. *Ambio, 50*, 834–869. https://doi.org/10.1007/s13280-021-01544-8

Frantzeskaki, N., Lambertucci, S., Carr, E., Dempsey, J., Karamidehkordi, E., Sengupta, A., Rojas Marchini, F., Vogel, C., Andriamahefazafy, M., Boonstra, W., Espinoza-Cisneros, E., Garcia, K., Morita, K., Nelson, V., Ojeda, D., Plieninger, T., Stirling, A., Tokunaga, K., Chen, R., …, & Guibal, C. (2024). Overcoming the challenges of achieving transformative change towards a sustainable world. In K. O'Brien, L. Garibaldi, & A. Agrawal (Eds.), *Thematic assessment report on the underlying causes of biodiversity loss and the determinants of transformative change and options for achieving the 2050 Vision for Biodiversity*. IPBES secretariat. https://doi.org/10.5281/zenodo.11382246

Fünfgeld, H. (2015). Facilitating local climate change adaptation through transnational municipal networks. *Current Opinion in Environmental Sustainability, 12*, 67–73. https://doi.org/10.1016/j.cosust.2014.10.011

Gasser, T., Guivarch, C., Tachiiri, K., Jones, C. D., & Ciais, P. (2015). Negative emissions physically needed to keep global warming below 2°C. *Nature Communications, 6*(1), 7958. https://doi.org/10.1038/ncomms8958

Gazing Wolf, J., Ignace, D. D., David-Chavez, D. M., Jennings, L. L., Smiles, D., Blanchard, P., Simmons, E., Doan-Crider, D., Kills, R.P.S.-S., Montgomery, M., Nelson, M. K., Black Elk, L., Black Elk, L., Bridge, G., Chischilly, A. M., Deer, K., DeerinWater, K., Ecoffey, T., Vergun, J., …, & Rattling Leaf, J. (2024). Centering indigenous knowledges in ecology and beyond. *Frontiers in Ecology and the Environment, 22*(7), Article e2776. https://doi.org/10.1002/fee.2776

GCoM. (2024). *The 2024 global covenant of mayors impact report*. https://www.globalcovenantofmayors.org/impact2024/

Grimm, N. B., Faeth, S. H., Golubiewski, N. E., Redman, C. L., Wu, J., Bai, X., & Briggs, J. M. (2015). Global change and the ecology of cities. *Science, 319*(5864), 756–760.

Gulsrud, N. M., Hertzog, K., & Shears, I. (2018). Innovative urban forestry governance in Melbourne? Investigating 'green placemaking' as a nature-based solution. *Environmental Research, 161*, 158–167. https://doi.org/10.1016/j.envres.2017.11.005

Hale, T. (2016). "All hands on deck": The Paris Agreement and nonstate climate action. *Global Environmental Politics, 16*(3), 12–22. https://doi.org/10.1162/GLEP_a_00362

Hickmann, T., Widerberg, O., Lederer, M., & Pattberg, P. (2021). The united nations framework convention on climate change secretariat as an orchestrator in global climate policymaking. *International Review of Administrative Sciences, 87*(1), 21–38. https://doi.org/10.1177/0020852319840425

Hollis, H., Skropke, C., Smith, H., Harries, H., & Garling, O. (2023). *Social infrastructure: International comparative review*. IFCS & Bennett IFPP. www.bennettinstitute.cam.ac.uk/wp-content/uploads/2023/01/Social_infrastructure_international_comparative_review.pdf

Houlden, V., Weich, S., Albuquerque, J. P., Jarvis, S., & Rees, K. (2018). The relationship between greenspace and the mental wellbeing of adults: A systematic review. *PLoS ONE, 13*(9), Article e0203000. https://doi.org/10.1371/journal.pone.0203000

Hunter, N., Roberts, D., Sutherland, C., & Slotow, R. (2022). Co-creation between cities and climate change science achieves research and action agenda. *Current Research in Environmental Sustainability, 4*, Article 100189. https://doi.org/10.1016/j.crsust.2022.100189

Hunter, N., Walsh, B., Delgado Ramos, G. C., Greenwalt, J., Roberts, D., Oke, C., Nielsen, D., & Slotow, R. (2023). Cities and climate change science after Edmonton: Knowledge, people, and organisations' influence and outreach. *Current Research in Environmental Sustainability, 6*, Article 100231. https://doi.org/10.1016/j.crsust.2023.100231

Hurlimann, A., Cobbinah, P. B., Bush, J., & March, A. (2020). Is climate change in the curriculum? An analysis of Australian urban planning degrees. *Environmental Education Research, 26*(9–10), 1349–1365. https://doi.org/10.1080/13504622.2020.1836132

ICLEI. (2024). *CHAMP: Coalition for high ambition multilevel partnership for climate action*. https://iclei.org/champ/

IEA. (2025). *Global energy review 2025*. https://iea.blob.core.windows.net/assets/5b169aa1-bc88-4c96-b828-aaa50406ba80/GlobalEnergyReview2025.pdf

IPBES. (2019). *Global assessment report on biodiversity and ecosystem services of the Intergovernmental Science-Policy Platform on Biodiversity and Ecosystem Services*. IPBES Secretariat. https://ipbes.net/global-assessment-report-biodiversity-ecosystem-services

IPCC. (2016). *Report of the forty-third session of the IPCC*. Nairobi, Kenya, 11–13 April 2016. https://www.ipcc.ch/site/assets/uploads/2018/05/final_report_p43.pdf

IPCC. (2022). *Climate change 2022: Mitigation of climate change. Contribution of working group III to the sixth assessment report of the intergovernmental panel on climate change*, Chapter 8: Urban systems and other settlements. Cambridge University Press. Available at: https://www.ipcc.ch/report/ar6/wg3/. Accessed 27 November 2025. IPCC+1

IPCC. (2024). *Decisions adopted by the panel (Decision IPCC-LXI-5)*. 61st session. https://www.ipcc.ch/site/assets/uploads/2024/08/IPCC-61_decisions-adopted-by-the-Panel.pdf

IRP-UNEP. (2018). *The weight of cities: Resource requirements of future urbanization*. www.resourcepanel.org/file/971/download?token=ehOygAQ7

Irvin, A., Oke, C., Khan, S., Jance IV, B., Ranalder, L., Barth, B., Deacon, A., Abusamra, R., Addison, A., Adebisi, J., Adsett, C., Ailio, E., Al Jadaa, D., Aslan, S., Assalini, S., Assarkhaniki, Z., Astolfo, J. P., Badino, M., Bai, X.,

..., & Williams, A. (2024). Next-generation city climate action: Findings from the 2024 Innovate4Cities conference and update to the global research and action agenda on cities and climate change science (GRAA). Global Covenant of Mayors for Climate and Energy, UN-Habitat & University of Melbourne. https://doi.org/10.26188/27147612

Ives, C. D., & Baker, C. (2024). Engaging faith for a sustainable urban future. *Global Sustainability, 7*, Article e45. https://doi.org/10.1017/sus.2024.32

Jance, B., Irvin, A., Oke, C., Assalini, S., Barth, B., Deacon, A., Jones, A., Jones-Langley, J., Khan, S., Mariani, P., Melica, G., Oussidhoum, F., Ranalder, L., Parry, L., Salehi P., & Tacconi, M. (2024). *Shared spaces for urban innovation: Insights and outcomes from Innovate4Cities marketplaces.* Global Covenant of Mayors for Climate & Energy. https://www.globalcovenantofmayors.org/wp-content/uploads/2024/02/I4C-Marketplaces-Outcomes-Insights-Report_.pdf

Joshi, A., & Moore, M. (2004). Institutionalised co-production: Unorthodox public service delivery in challenging environments. *Journal of Development Studies, 40*(4), 31–49.

Kabisch, N., Basu, S., van den Bosch, M., Bratman, G. N., & Masztalerz, O. (2023). Nature-based solutions and mental health. In T. McPhearson, N. Kabisch, & N. Frantzeskaki (Eds.), *Nature-based solutions for cities* (pp. 192–212). Edward Elgar Publishing. https://doi.org/10.4337/9781800376762.00019

Kapucu, N., Ge, Y., Rott, E., & Isgandar, H. (2024). Urban resilience: Multidimensional perspectives, challenges and prospects for future research. *Urban Governance Journal, 2*(3), 245–268. https://doi.org/10.1016/j.ugj.2024

Kendal, D., Egerer, M., Byrne, J. A., Jones, P. J., Marsh, P., Threlfall, C. G., Allegretto, G., Kaplan, H., Nguyen, H. K. D., Pearson, S., Wright, A., & Flies, E. J. (2020). City-size bias in knowledge on the effects of urban nature on people and biodiversity. *Environmental Research Letters, 15*(12), Article 124035. https://doi.org/10.1088/1748-9326/abc5e4

Kirchengast, G., & Pichler, M. (2025). A traceable global warming record and clarity for the 1.5 °C and well-below-2 °C goals. *Communications Earth & Environment*, *6*, 402. https://doi.org/10.1038/s43247-025-02368-0

Kitchin, R., Lauriault, T. P., & McArdle, G. (2015). Smart cities and the politics of urban data. In *Smart urbanism* (pp. 16–33). Routledge.

Langemeyer, J., & Baró, F. (2021). Nature-based solutions as nodes of green-blue infrastructure networks: A cross-scale, co-creation approach. *Nature-Based Solutions, 1*, Article 100006. https://doi.org/10.1016/j.nbsj.2021.100006

Li, Y., Zhang, Y., & Yu, X. (2019). Urban weight and its driving forces: A case study of Beijing. *Science of the Total Environment, 658*, 590–601. https://doi.org/10.1016/j.scitoenv.2018.12.022

Lwasa, S., Seto, K. C., Bai, X., Blanco, H., Gurney, K. R., Kılkış, Ş., Lucon, O., Murakami, J., Pan, J., Sharifi, A., & Yamagata, Y. (2022). Urban systems and other settlements. In *IPCC, 2022: Climate change 2022: Mitigation of climate change. Contribution of working group III to the sixth assessment report of the intergovernmental panel on climate change*. Cambridge University Press. https://doi.org/10.1017/9781009157926.010

Mahadevia, D., Delgado Ramos, G. C., Barnes, J., Fitzgerald, J., Kamei, M., & Lanza, K. (2025). Learning from COVID-19 for climate-ready urban transformation. *Cambridge University Press*. https://doi.org/10.1017/9781009527279

McPhearson, T., Pickett, S. T. A., Grimm, N. B., Niemelä, J., Alberti, M., Elmqvist, T., Weber, C., Haase, D., Breuste, J., & Qureshi, S. (2016). Advancing urban ecology toward a science of cities. *BioScience, 66*(3), 198–212. https://doi.org/10.1093/biosci/biw002

Medeiros, E., & van der Zwet, A. (2020). Sustainable and integrated urban planning and governance in metropolitan and medium-sized cities. *Sustainability, 12*(15), 5976.

Moreno, C., Allam, Z., Chabaud, D., Gall, C., & Pratlong, F. (2021). Introducing the "15-Minute City": Sustainability, resilience and place identity in future post-pandemic cities. *Smart Cities, 4*(1), 93–111. https://doi.org/10.3390/smartcities4010006

Municipality of Lima. (2024). *Air quality monitoring network expansion report*. https://www.munlima.gob.pe/

Nelson, M. K., & Shilling, D. (Eds.). (2018). *Traditional ecological knowledge: Learning from indigenous practices for environmental sustainability*. Cambridge University Press.

Oke, C., Walsh, B., Assarkhaniki, Z., Jance, B., Deacon, A., & Lundberg, K. (2022). The city research and innovation agenda: Prioritizing knowledge gaps and policy processes to accelerate city climate action. *Journal of City Climate Policy and Economy, 1*(1), 94–110. https://doi.org/10.3138/jccpe-2022.1.1.0007

Orlove, B., Sherpa, P., Dawson, N., Adelekan, I., Alangui, W., Carmona, R., Coen, D., Nelson, M. K., Reyes-García, V., Rubis, J., Sanago, G., & Wilson, A. (2023). Placing diverse knowledge systems at the core of transformative climate research. *Ambio, 52*(9), 1431–1447. https://doi.org/10.1007/s13280-023-01857-w

Pereira, L. M., Davies, K. K., den Belder, E., Ferrier, S., Karlsson-Vinkhuyzen, S., Kim, H., Kuiper, J. J., Okayasu, S., Palomo, M. G., Pereira, H. M., Peterson, G., Sathyapalan, J., Schoolenberg, M., Alkemade, R., Carvalho Ribeiro, S., Greenaway, A., Hauck, J., King, N., Lazarova, T., …, & Lundquist, C. J. (2020). Developing multiscale and integrative nature–people scenarios using the Nature Futures Framework. *People and Nature, 2*(4), 1172–1195. https://doi.org/10.1002/pan3.10146

Pörtner, H. O., Scholes, R. J., Agard, J., Archer, E., Arneth, A., Bai, X., Barnes, D., Burrows, M., Chan, L., Cheung, W. L., Diamond, S., Donatti, C., Duarte, C., Eisenhauer, N., Foden, W., Gasalla, M. A., Handa, C., Hickler, T., Hoegh-Guldberg, O., ... & Ngo, H. T. (2021). *IPBES-IPCC co-sponsored workshop report on biodiversity and climate change. IPCC and IPBES.* https://www.ipbes.net/sites/default/files/2021-06/20210609_workshop_report_embargo_3pm_CEST_10_june_0.pdf

Prieur-Richard, A.-H., Walsh, B., Craig, M., Melamed, M. L., Colbert, M., Pathak, M., Connors, S., Bai, X., Barau, A., Bulkeley, H., Cleugh, H., Cohen, M., Colenbrander, S., Dodman, D., Dhakal, S., Dawson, R., Espey, J., Greenwalt, J., Kurian, P., …, & Ürge-Vorsatz, D. (2019). *Global research and action agenda on cities and climate change science.* Intergovernmental Panel on Climate Change. https://www.ipcc.ch/site/assets/uploads/2019/07/Research-Agenda-Aug-10_Final_Long-version.pdf

Quezon City Government. (2022). *GrowQC Urban agriculture initiative report.* https://quezoncity.gov.ph/

Ranalder, L., Tollin, N., Gragnani, P. & Geoghegan, P. (2025) *Seizing the urban opportunity: A 3-step guide for strengthening nationally determined contributions (NDC 3.0).* United Nations Human Settlements Programme (UN-Habitat). https://unhabitat.org/seizing-the-urban-opportunity-a-3-step-guide-for-strengthening-nationally-determined-contributions

Reisinger, A., Fuglestvedt, J. S., Pirani, A., Geden, O., Jones, C. D., Maharaj, S., Poloczanska, E. S., Morelli, A., Johansen, T. G., Adler, C., Betts, R. A., & Seneviratne, S. I. (2025). Overshoot: A conceptual review of exceeding and returning to global warming of 1.5 °C. *Annual Review of Environment and Resources, 50*, 185–217. https://doi.org/10.1146/annurev-environ-111523-102029

Renner, A. T., Plank, L., & Getzner, M. (2024). *Handbook of social infrastructure: Conceptual and empirical research perspectives*. Edward Elgar Publishing.

Revi, A., Babiker, M., Bertoldi, P., Cartwright, A., Chow, W., Colenbrander, S., Dawson, R., De Coninck, H., De Kleijne, K., Dhakal, S., Gallardo, L., Garschagen, M., Haasnoot, M., Haldar, S., Hamdi, R., Hashizume, M., Islam, A. K. M. S., Jiang, K., Kılkış, S., ... & Ürge-Vorsatz, D. (2022). *The summary for urban policymakers of the IPCC's sixth assessment report*. Indian Institute for Human Settlements. https://doi.org/10.24943/SUPSV511.2022

Reynolds, H. L., Brandt, L., Fischer, B. C., Hardiman, B. S., Moxley, D. J., Sandweiss, E., Speer, J. H., & Fei, S. (2020). Implications of climate change for managing urban green infrastructure: An Indiana, US case study. *Climatic Change, 163*, 1967–1984. https://doi.org/10.1007/s10584-019-02617-0

Ricciardi, G., & Callegari, G. (2022). Digital twins for climate-neutral and resilient cities. State of the art and future development as tools to support urban decision-making. In *International conference on technological imagination in the green and digital transition* (pp. 617–626). Springer International Publishing.

Rosenzweig, C., Solecki, W. D., Romero-Lankao, P., Mehrotra, S., Dhakal, S., & Ibrahim, S. A. (Eds.). (2018). *Climate change and cities: Second assessment report of the urban climate change research network.* Cambridge University Press.

Ruini, A., Sporchia, F., Niccoluci, V., Pulselli, F. M., & Bastianoni, S. (2025). Rethinking environmental benefit allocation in industrial symbiosis. *Science of the Total Environment, 992*, Article 179932. https://doi.org/10.1016/j.scitotenv.2025.179932

Seto, K. C., Dhakal, S., Bigio, A., Blanco, H., Delgado, G. C., Dewar, D., Huang, L., Inaba, A., Kaneko, A., Lwasa, S., McMahon, J. E., Müller, D. B., Murakami, J., Nagendra, H., & Ramaswami, A. (2014). Human settlements, infrastructure and spatial planning. In *Climate change 2014: Mitigation of climate change* (pp. 923–1000). Cambridge University Press.

Shand, W. (2018). Making spaces for co-production: Collaborative action for settlement upgrading in Harare, Zimbabwe. *Environment and Urbanization, 30*(2), 519–536.

Shi, Y., Sorrell, S., & Foxon, T. (2023). The impact of teleworking on domestic energy use and carbon emissions: An assessment for England. *Energy and Buildings, 287*, Article 112996. https://doi.org/10.1016/j.enbuild.2023.112996

Simon, S., & O'Brien, W. (2023). Pilot study to measure the energy and carbon impacts of teleworking. *Buildings & Cities, 4*(1), 174–192. https://doi.org/10.5334/bc.271

Solecki, W., Griffith, C., Larsen, L., Revi, A., Romero-Lankao, P., Smit, W., Fragkias, L., Anguelovski, E., Arslan, U., Barbieri, R., Bazán, L. M., Cobbinah, P., De Sherbinin, A., Elmqvist, T., Lwasa, S., Mende, B., Mulugetta, Y., Naess, K., Nilsson, M., ..., & Kernaghan, A. (2021). Growing exclusive knowledge inclusive of cities: From historical impacts to future pathways. *Current Research in Environmental Sustainability, 3*, Article 100038. https://doi.org/10.1016/j.crsust.2021.100038

Tengö, M., Austin, B. J., Danielsen, F., & Fernández-Llamazares, A. (2021). Creating synergies between citizen science and Indigenous and local knowledge. *BioScience, 71*(5), 503–518. https://doi.org/10.1093/biosci/biab023

Theokritoff, E., Lejeune, Q., Costa, H. P., Irfan, K., Khan, M. S., Kropf, C. M., Lindberg, H. G., Marques, I. G., Menke, I., Schleussner, C. F., Thomas,

A., & Lourenço, T. C. (2025). Climate overshoot implications for local adaptation planning. *Climate Policy*, 1-8. https://doi.org/10.1080/14693062.2025.2502111

Tollin, N., Vener, J., Liu, Y., Gragnani, P., Pizzorni, M., & Barth, B. (2024). *Urban content of NDCs: Local climate action explored through in-depth country analysis 2024 report*. UN-Habitat. https://unhabitat.org/sites/default/files/2025/09/ndc_global_report_2024.pdf

UN DESA. (2018). *World urbanization prospects: The 2018 revision (ST/ESA/SER.A/420)*. United Nations Department of Economic and Social Affairs, Population Division.

UN DESA. (2024). *Policy Brief No. 162: Multilevel governance for climate change mitigation and adaptation*. https://desapublications.un.org/policy-briefs/un-desa-policy-brief-no-162-multilevel-governance-climate-change-mitigation-and

UN-Habitat. (2018). *Cities and climate change science conference proceedings*. https://unhabitat.org/sites/default/files/2021/03/Cities%20IPCC%20Proceedings%20FINAL%20for%20Email-S.pdf

UN-Habitat. (2024). *World cities report 2024: Cities and climate action*. United Nations Human Settlements Programme.

UN-Habitat. (2025a). *Seizing the urban opportunity: A 3-step guide for strengthening nationally determined contributions 3.0*. https://unhabitat.org/sites/default/files/2025/06/3-step-guide-for-strengthening-ndcs.pdf

UN-Habitat. (2025b). *Urban content in NDC 3.0: A global snapshot for COP30*. United Nations Human Settlements Programme. Available at: https://unhabitat.org. Accessed 25 November 2025.

UNEP-UN-Habitat. (2021). *Global environment for cities-GEO for cities: Towards green and just cities*. UNEP. https://wedocs.unep.org/bitstream/handle/20.500.11822/37413/GEOcities.pdf

UNESCAP. (2020). *The future of Asian and Pacific Cities Report 2022: Building resilient cities beyond the COVID-19 pandemic*. United Nations Economic and Social Commission for Asia and the Pacific. https://www.unescap.org/sites/default/d8files/knowledge-products/The%20Future%20of%20Asian%20%26%20Pacific%20Cities_FINAL%2827%20Jan%2021%29.pdf

United Nations Framework Convention on Climate Change (UNFCCC). (2025). *2025 NDC synthesis report*. UNFCCC. Available at: https://unfccc.int/documents/650664. Accessed 27 November 2025.

Van Nes, A. (2021). Spatial configurations and walkability potentials: Measuring urban compactness with space syntax. *Sustainability, 13*(11), 5785. https://doi.org/10.3390/su13115785

Vogel, E., Davis, K. J., Wu, K., Miles, N. L., Richardson, S. J., Gurney, K. R., Monteiro, V., Roest, G. S., Kenion, H. C. R., & Horne, J. P. (2024). Using eddy-covariance to measure the effects of COVID-19 restrictions on CO_2

emissions in a neighborhood of Indianapolis. *Carbon Management, 15*(1), 2365900. https://doi.org/10.1080/17583004.2024.2365900

Walsh, B., Greenwalt, J., Oke, C., Hadfield, P., Dickey, A., H Craig, M., North, M., Hunter, N., Okem, A., Bland, S., Verbeeck, J., Barth, B., Deacon, A., Roberts, H., Randhawa, J., Joshi, S., Silva, D. C., Lootens, M., Adler, M., …, & Zhu, S. (2022). *Findings from Innovate4Cities 2021 and update to the global research and action agenda (Version 3).* The University of Melbourne. https://doi.org/10.26188/19624029.v3

Waterton, E. (2015). Heritage and community engagement. In Ireland, T., & Schofield, J. (Eds.), *The ethics of cultural heritage* (pp. 53–67). Springer. https://doi.org/10.1007/978-1-4939-1649-8_4

WHO. (2025, March 19). *Urban health.* https://www.who.int/news-room/fact-sheets/detail/urban-health

Williams, J. (2022). Circular cities: Planning for circular development in European cities. *European Planning Studies, 31*(1), 14–35. https://doi.org/10.1080/09654313.2022.2060707

World Bank & UNCDF. (2024). *Local government climate finance instruments: Global experience and prospects in developing countries.* https://documents1.worldbank.org/curated/en/099041224090039327/pdf/P17612810871700531ae091b1c089264caf.pdf

WRI. (2025). *Climate watch insights: Trends in the 2025 NDC submissions.* World Resources Institute. Available at: https://www.climatewatchdata.org. Accessed 25 November 2025.

Yeh, S.-C., Chen, Y.-H., Van Velzen, R., & Lin, P.-H. (2025). The climate change literacy of public officials in Taiwan: Implications and strategies for global adaptation. *Local Government Studies, 51*(2), 234–256. https://doi.org/10.1080/01442872.2024.2304086

Index

C. Oke et al. (eds.), *A Global Research and Action Agenda on Cities and Climate Change Science*,
https://doi.org/10.1007/978-981-95-6517-7

H

I

J